5th International Conference on New Material and Chemical Industry (NMCI 2020)

Held online due to COVID-19

IOP Conference Series: Materials Science and Engineering
Volume 1040

Xiamen, China
14 - 16 November 2020

Editor:

Shixuan Xin

ISBN: 978-1-7138-3330-7
ISSN: 1757-8981

Printed from e-media with permission by:

Curran Associates, Inc.
57 Morehouse Lane
Red Hook, NY 12571

Some format issues inherent in the e-media version may also appear in this print version.

Additional copies of this publication are available from:

Curran Associates, Inc.
57 Morehouse Lane
Red Hook, NY 12571 USA
Phone: 845-758-0400
Fax: 845-758-2633
Email: curran@proceedings.com
Web: www.proceedings.com

TABLE OF CONTENTS

Preface

Shixuan Xin

Editor, Proceedings of The Fifth International Conference on New Material and Chemical Industry (NMCI 2020)

Professor, Petro China Company Limited, China

Email: xinshixuan@petrochina.com.cn

The Fifth International Conference on New Material and Chemical Industry (NMCI 2020) was planned to be held on November 14-16, 2020 in Xiamen, China. Due to the COVID-19 epidemic and to avoid the risk of crowd gathering, the organizing committee decided to hold NMCI 2020 virtually/online on November 14-16, 2020. Although held online, this year's conference also attracted many scholars and experts in the field of New Material and Chemical Industry to share new research results, novel policy developments, and practical implementation experiences. NMCI2020 was a successful, fruitful online academic event. Nearly 30 participants from 3 different countries and regions attended the online meeting.

The conference consists of three parts: keynote presentations, oral presentations and poster presentations and discussions covering a wide range of subjects from New non-ferrous metal alloy, Nano materials and technology, New building materials, New chemical material, Organic chemical industry to Chemical engineering. We are very grateful that Professor Shixuan Xin from Petro China Company Limited, China presided over the conference. Thanks for all participants' attending and sharing!

It is our great honor to invite 5 experts to give the outstanding keynote speeches: Prof. Laichang Zhang, from School of Engineering Edith Cowan University; Prof. Zhenyu Tian, from Chinese Academy of Sciences; Prof. Donatella Giuranno, from National Research Council of Italy, Prof. Xin Jin, from China University of Petroleum, China and Dr. Jie Wei from University of Electronic Science and Technology of China.

After keynote presentations, there were 3 orals presentations and 4 poster presentations. Oral speakers and poster presenters shared their latest research results with the audience, the audience actively and enthusiastically participated in by asking questions and voting.

This conference proceeding included 18 accepted articles selected from 45 submissions.

We would like to express our gratitude to the reviewers of these manuscripts, who provided their time and expertise to review papers and facilitate the smooth running of NMCI. We are extremely grateful for technical program committee and editors and extend our most sincere thanks to all the authors for their excellent contribution and work and participants for their active attending. Our sincere thanks also go to the IOP Publishing editors and managers for their helpful cooperation during the preparation of the proceeding.

On behalf of the Organizing Committee of NMCI 2020

IOP Conf. Series: Materials Science and Engineering 1040 (2021) 011001 doi:10.1088/1757-899X/1040/1/011001

Technical Program Committee

Conference Chair
Shixuan Xin, China Company Limited, China
Conference Co-chair
Xiaorong Luo, University of Electronic Science and Technology of China, China
TPC Chair
Manuel F. M. Costa, University of Minho, Portugal

TPC member
Jincang Zhang, Shanghai Univerisity, China
Abdeltif Amrane, University of Rennes 1, France
Jun Li, Kansas State University, USA
Weidong Wang, Xidian University, China
Meng Gu, Southern University of Science and Technology, China
Catalin Iulian Pruncu, Imperial College London, UK
An Liang, The Hong Kong Polytechnic University, China
Bakshi Hardeep Vaid, Nanjing University of Information Science and Technology
Dai-Viet N. Vo, Nguyen Tat Thanh University, Vietnam
Mohammad Sadiq, University of Malakand, Pakistan
Larisa G. Gordeeva, Boreskov Institute of Catalysis, Russia
Alexander Tolstov, Institute of Macromolecular Chemistry of the NAS of Ukraine, Ukraine
Yasin Polat, Nevşehir Hacı Bektaş Veli University, Turkey
Kheng-Lim Goh, Newcastle University, Singapore

IOP Conf. Series: Materials Science and Engineering 1040 (2021) 011001 doi:10.1088/1757-899X/1040/1/011001

Conference Overview

Due to COVID-19 epidemic, The Fifth International Conference on New Material and Chemical Industry (NMCI 2020) was held online on November 14-16, 2020. The conference aims to provide the best platform for researchers and scholars worldwide to discuss recent researches and developments in the area of New Material and Chemical Industry.

Below is the overview of NMCI2020 online conference.

Nearly 30 participants from 3 different countries and regions attended the online meeting. The conference consists of three parts: keynote presentations, oral presentations and poster presentations. Professor Shixuan Xin from Petro China Company Limited, China, presided over the conference.

There are 5 keynote speeches delivered in this conference.

Figure 1. Keynote Presentation Video from Prof. Laichang Zhang.

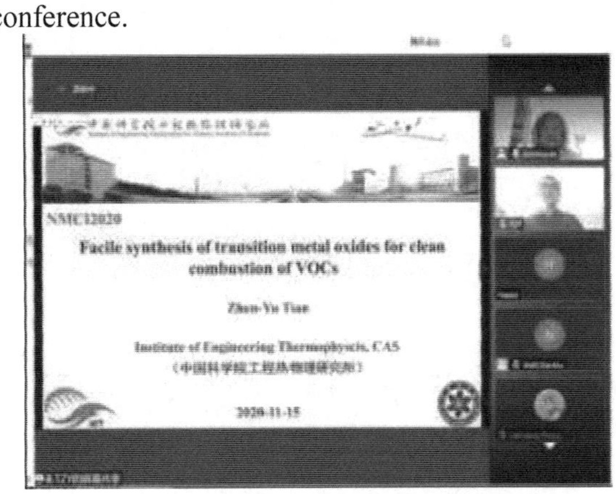

Figure 2. Keynote Presentation from Prof. Zhenyu Tian.

IOP Conf. Series: Materials Science and Engineering 1040 (2021) 011001 doi:10.1088/1757-899X/1040/1/011001

Prof. Laichang Zhang, from School of Engineering Edith Cowan University, firstly made the presentation on “Mechanical behavior of 3D printed titanium alloy porous scaffolds”**(Figure 1)**.

The second speech was delivered by Prof. Zhenyu Tian, from Chinese Academy of Sciences. The theme of the speech is “Facile synthesis of transition metal oxides for clean combustion of VOCs” **(Figure 2)**.

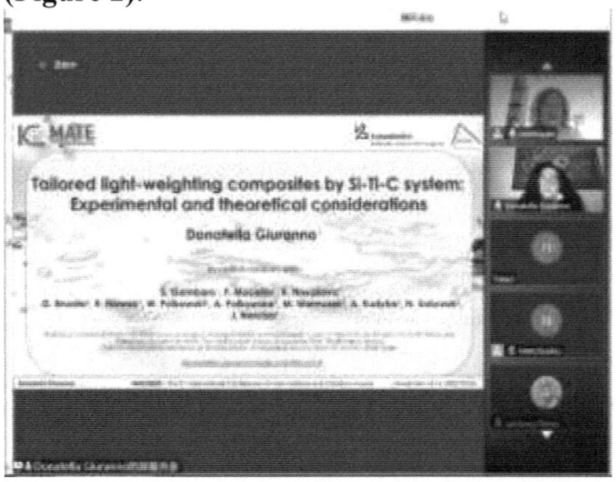

Figure 3. Keynote Presentation from Prof. Donatella Giuranno.

Figure 4. Keynote Presentation from Prof. Xin Jin.

Prof. Donatella Giuranno, from National Research Council of Italy gave a wonderful speech on “Tailored light-weighting composites by Si-Ti-C system: Experimental and theoretical considerations” (**Figure 3**).

Prof. Xin Jin, from China University of Petroleum, China gave a presentation on “Catalytic conversion of sugars to dicarboxylic acids over bimetallic catalysts” (**Figure 4**).

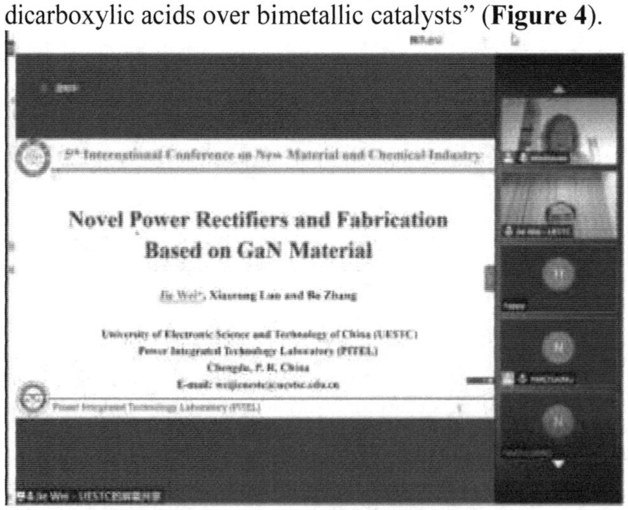

Figure 5. Keynote Presentation from Dr. Jie Wei.

As the last keynote speaker, Dr. Jie Wei, from University of Electronic Science and Technology of China gave a presentation on “Novel Power Rectifiers and Fabrication Based on GaN Material” (**Figure 5**).

After keynote presentations, there are 3 oral presentations (**Figure 6**) and 4 poster presentations (**Figure 7**). Oral speakers and poster presenters shared their latest research results with the audience, the audience actively and enthusiastically participated in by asking questions and online voting.

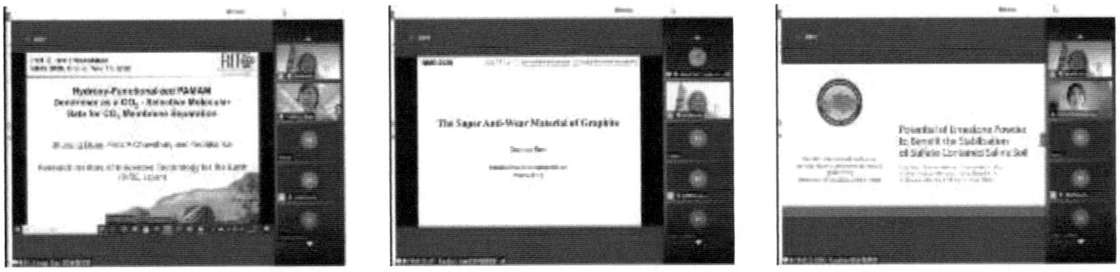

Figure 6. 3 Oral Presentations of NMCI2020.

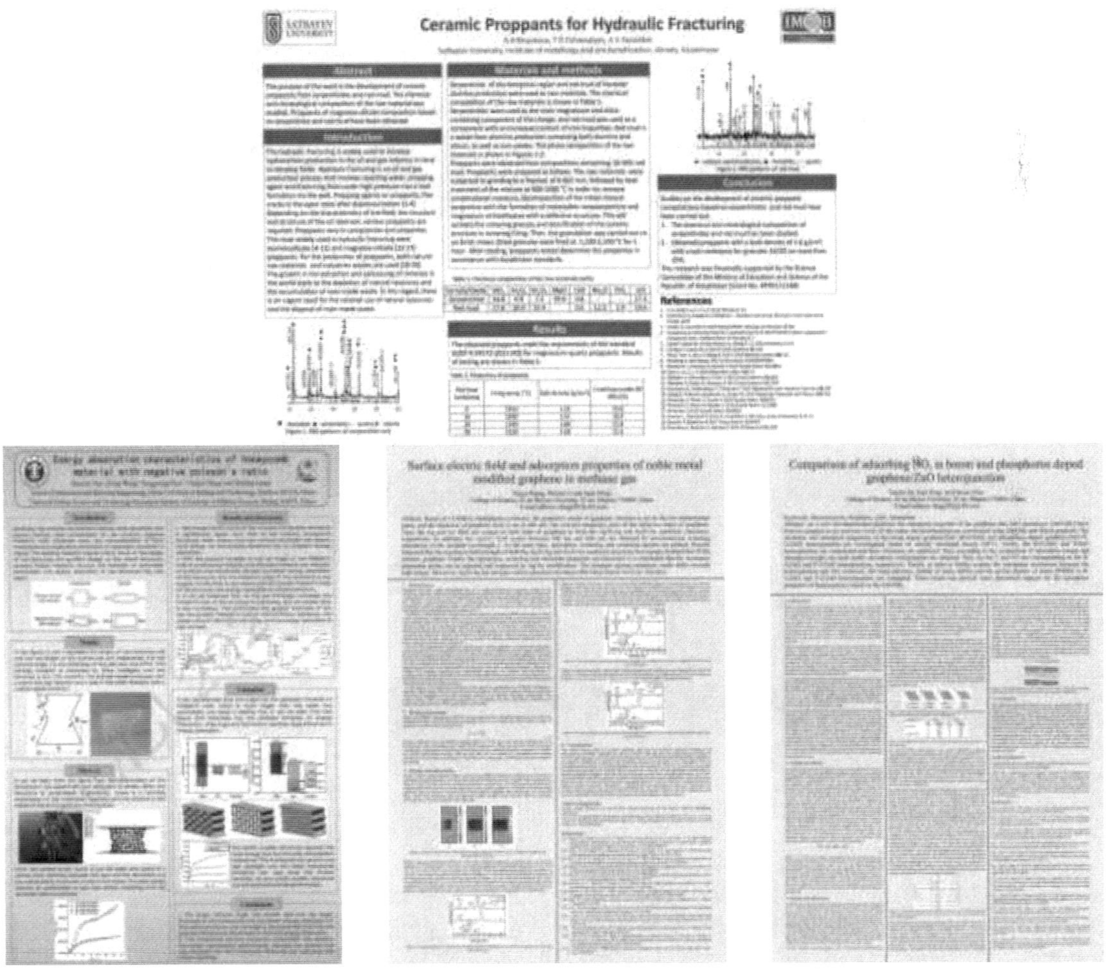

Figure 7. 4 Poster Presentations of NMCI2020.

The conference was a complete success, thanks to the support of the participants, especially the host, keynote speakers and oral speakers. We are expecting more and more experts and scholars from around the world to join this international conference next year.

IOP Conf. Series: Materials Science and Engineering 1040 (2021) 011002 doi:10.1088/1757-899X/1040/1/011002

Peer review declaration

All papers published in this volume of *IOP Conference Series: Materials Science and Engineering* have been peer reviewed through processes administered by the Editors. Reviews were conducted by expert referees to the professional and scientific standards expected of a proceedings journal published by IOP Publishing.

- Type of peer review: Double-blind

Author' s information were removed from the papers before be forwarded to the reviewers.

And the authors only got the comments of papers without reviewers' information.

- Conference submission management system:

http://papersub.icnmci.org/

- Number of submissions received:

45

- Number of submissions sent for review:

38

- Number of submissions accepted:

18

- Acceptance Rate (Number of Submissions Accepted / Number of Submissions

Received X 100):

Approximately 40%

- Average number of reviews per paper:

3

- Total number of reviewers involved:

144

- Any additional info on review process:

After preview by editorial assistant, 45 papers were sent for peer reviews, 144 reviewers involved in the process, and review process for each paper took around 1 month. The decision to accept or reject was based on the reviewers' comments. If differences of opinion occurred between reviewers, the editors weighed all comments and arrived at a balanced decision based on all comments, or a second round of peer review was initiated.

- Contact person for queries:

Ms.Zheng, sec_zheng@icnmci.org

Synthesis of submicron-sized carbon-doped TiO_2 for photodegradation of formaldehyde from wood-based panels

Ligang Luo[1], Weihui Yao[1], Ronghui Yuan[1], Wei Shi[1], Xiaohong Jin[1], Jianhua Hu[2] and Guojun Jiang[3,4]

[1]Zhejiang Light Industrial Product Inspection and Research Institute, Hangzhou 310018, China;
[2]Hangzhou Wanxiang Polytechnic, Hangzhou 310023, China;
[3]Zhijiang College of Zhejiang University of Technology, Shaoxing 312030, China

[4]Email: jiangguojun1986@126.com

Abstract. Carbon-doped TiO_2 particles were prepared by hydrothermal treatments and characterised via field emission scanning electron microscopy, powder X-ray diffraction and X-ray photoelectron spectroscopy. The results show that carbon was successfully doped into the TiO_2 structures. Photocatalytic reactions were further investigated in a simulated indoor environment, and the visible-light catalytic activity of the carbon-doped TiO_2 was evaluated by monitoring the photodegradation of the formaldehyde from wood-based panels. A degradation rate of 90% in 70 min was obtained, which demonstrated high photocatalytic performance.

1. Introduction

As listed in the category of carcinogens by the World Health Organization, formaldehyde has been reported to cause issues in the nervous and respiratory systems [1-3]. Nowadays, people are more concerned about the health implication of wood-based panels, especially the indoor-used panels, which usually emit excessive formaldehyde [4]. Adhesives play important roles in the production process of wood-based panels (widely used in the market), and they are mainly used to combine panels and other attachments [5]. Formaldehyde exists as a volatile compound in adhesives [6]. It is obtained from urea formaldehyde and phenol aldehyde resins. With strong adhesiveness, it increases the hardness of the board, and it also repels insects and prevents corrosion. Formaldehyde gradually diffuses to the surrounding as a source of pollution in indoor environments. Therefore, there is a need to investigate wood-based panels, including blockboard, particleboard, medium-density fibreboard, veneer and plywood, produced with adhesives to determine the formaldehyde emissions and minimise the possible harmful effect.

TiO_2 is known for its low cost, chemical and thermal stability, resistance to photo corrosion and nontoxicity [7,8]. Photocatalysts are developed for more effective utilisation of the natural light [9]. TiO_2 modified with non-metals has been extensively employed in formaldehyde degradation due to its outstanding performance [10-12]. Moreover, the photocatalytic activities of carbon-doped TiO_2 have attracted much interest owing to their potential in the removal of organic pollutants [13-15]. Thus, it is necessary to investigate visible-light active catalysts for formaldehyde degradation in a simulated indoor environment.

For this reason, submicron-sized carbon-doped TiO_2 particles were prepared via a hydrothermal method in this study, and their morphology and properties were characterised via field emission scanning electron microscopy (FE-SEM), powder X-ray diffraction (XRD), energy dispersive X-ray (EDX) and X-ray photoelectron spectroscopy (XPS). Simulated with 120-L environmental chambers, the formaldehyde emissions from wood-based panels were degraded in chambers under visible-light irradiation. The activity of carbon-doped TiO_2 was evaluated via photocatalytic tests, and the formaldehyde degradation was improved.

2. Materials and methods

2.1. Instruments and chemicals

The instruments used included a PNAlytical X Pert Pro multifunction XRD (Netherlands), FEI Quanta FEG 650 field-emission scanning electron microscope (Netherlands), Thermo Scientific EscaLab 250Xi XPS (USA) and Agilent Cary 300 UV-Vis spectrophotometer (USA). An Eagle View Electronic Technology 80 W LED lamp (China), Markes International TC-20 multi-tube ageing set (UK), Talboys magnetic hotplate stirrer (USA) and 120-L Y40D3 environmental test chamber (China) were also used.

Titanium butoxide (99%), D-glucose (98%), ammonium hydroxide (25%), acetylacetone and ethanol of at least the analytical grade were purchased from Aladdin (China). Acetylacetone (0.25%) was obtained by treating with a certain proportion of ammonium acetate/water/glacial acetic acid/acetylacetone media. 1000 mg L^{-1} standard solution of formaldehyde was purchased from TanMo (China).

2.2. Synthesis and characterisation of carbon-doped TiO_2

100 mL of anhydrous ethanol, 2.7 g of D-glucose and 16.4 mL of titanium butoxide were successively added to a beaker under stirring. The well-stirred mixture was added to a solution of ammonium hydroxide (0.5 mL) and ethanol (150 mL) using ultrasonic treatment for 10 min. Next, the resulting solution was transferred to a PTFE-lined hydrothermal reactor at 160 °C and stirred for 6 h. The precipitate was washed with ethanol and water, and then dried in an oven at 70 °C for 1 h. The precursors were calcined in a muffle furnace at 520 °C for 3 h. Thus, the carbon-doped TiO_2 photocatalyst was obtained. For a systematic study, the pure anatase TiO_2 was synthesised via a hydrothermal treatment [7].

The microstructure of carbon-doped TiO_2 was determined using the FE-SEM at an accelerating voltage of 20.0 kV. Gold (Au) was used as a coating material in the analysis of the plane scan. The XRD analysis was conducted using Cu-Kα radiation with a Cu X-ray tube operated at 30 kV. The interval of data points was 0.01°, and the scanning speed was 3° min^{-1}. The photocatalyst particles were scattered in ethanol by ultrasonic dispersion for 5 min, and then put on a copper mesh in a transmission electron microscope for EDX analysis. XPS spectra were obtained at a constant power of 150 W (binding energy: 100-700 eV) using monochromatic Al-Kα radiation.

2.3. Simulation conditions in environmental chambers

Prior to the operation of the chamber, its inner surface was scrubbed with alkaline detergent and rinsed with water. The simulation conditions of the chamber used for evaluation of formaldehyde levels from wood-based panels in indoor environments are listed in Table 1. A 5-level calibration curve was prepared in an aqueous solution with formaldehyde. The absorbance was in good agreement with the standard, and its correlation coefficient was 0.9993. The wood-based panels were cut to a size of 20 cm × 20 cm and sealed with aluminium foil tape at both the sides and the bottom surface (Figure 1*A*). Subsequently, the panels were placed in the chambers for testing (Figure 1*B*).

2.4. Photocatalytic activity measurement

A porous nickel foam net (20 cm × 20 cm × 1.6 mm) with 25 pores per linear inch was used as the catalyst carrier. 0.4 g carbon-doped TiO_2 photocatalysts was resuspended in 60 mL water and sonicated for 5 min. The suspension was sprayed to the net and placed under the LED lamp, thereafter, it was illuminated to start the reaction. Furthermore, 100 mL formaldehyde solution (initial concentration of 20 mg L^{-1}) was added to quartz containers. Before degradation, formaldehyde emissions were contacted with catalysts for 30 min to attain adsorption equilibrium. The formaldehyde degradation was then carried out at room temperature with the light source located at a certain distance from the nets. After the photodegradation, 5 mL extract liquor of large bubbling absorber (centrifuged at 2000 rpm for 5 min and filtered through a hydrophilic membrane) was uniformly mixed with 0.25% acetylacetone, and then heated in boiling water. The absorbance of the resulting solution was determined via UV-Vis spectrophotometry at a wavelength of 415 nm. The degradation rate of the formaldehyde (D_f, %) was calculated by the following equation:

$$D_f = \frac{A_0 - A_t}{A_0} \times 100\% \quad (1)$$

where A_0 and A_t are the absorbances of the solution before and after irradiation, respectively.

Table 1. Simulation conditions in environmental chambers.

Parameters	Conditions
Temperature	23 ± 1 °C
Relative humidity	50 ± 2%
Surface air velocity	0.1 m s^{-1}
Sampler	Large bubbling absorber
Air flow rate	0.1 L min^{-1}
Total sampling volume	2 L
Sample loading factor	0.33 m^2 m^{-3}
Visible-light wavelength	410 nm
Irradiation energy	76.3 mW cm^{-2}
Irradiation time	70 min

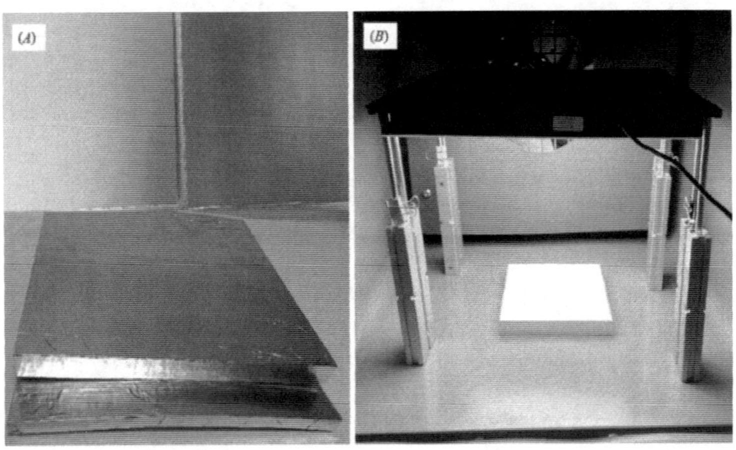

Figure 1. Schematic of aluminum foil tape preparation for wood-based panels (*A*) and photodegradation of BTEX in environmental chambers (*B*).

3. Results and discussion

3.1. Characterisations of carbon-doped TiO_2 photocatalysts

The carbon-doped TiO_2 particles appeared to be irregular shapes of submicron size (Figures 2*A* and 2*B*). The average grain size of the carbon-doped TiO_2 was measured to be 179 nm. The plane scans of micro-area for Ti and C are shown in Figures 2*C* and 2*D*. There was good compatibility between Ti and C during the calcination process, thus, a doped structure was obtained. Figure 2*E* indicates the XRD pattern of the carbon-doped TiO_2, and the product is a mixture consisting of anatase phase with the remaining small part being in (110) crystal plane of the rutile phase. The peaks at 25.3°, 36.9°, 37.8°, 38.6°, 48.0°, 53.9°, 55.1°, 62.7°, 68.8°, 70.3° and 75.1° were attributed to the diffraction of the (101), (103), (004), (112), (200), (105), (211), (204), (116), (220) and (215) crystal planes of the anatase phase, respectively. The EDX spectrum (Figure 2*F*) shows the chemical compositions of products are C, Ti and O.

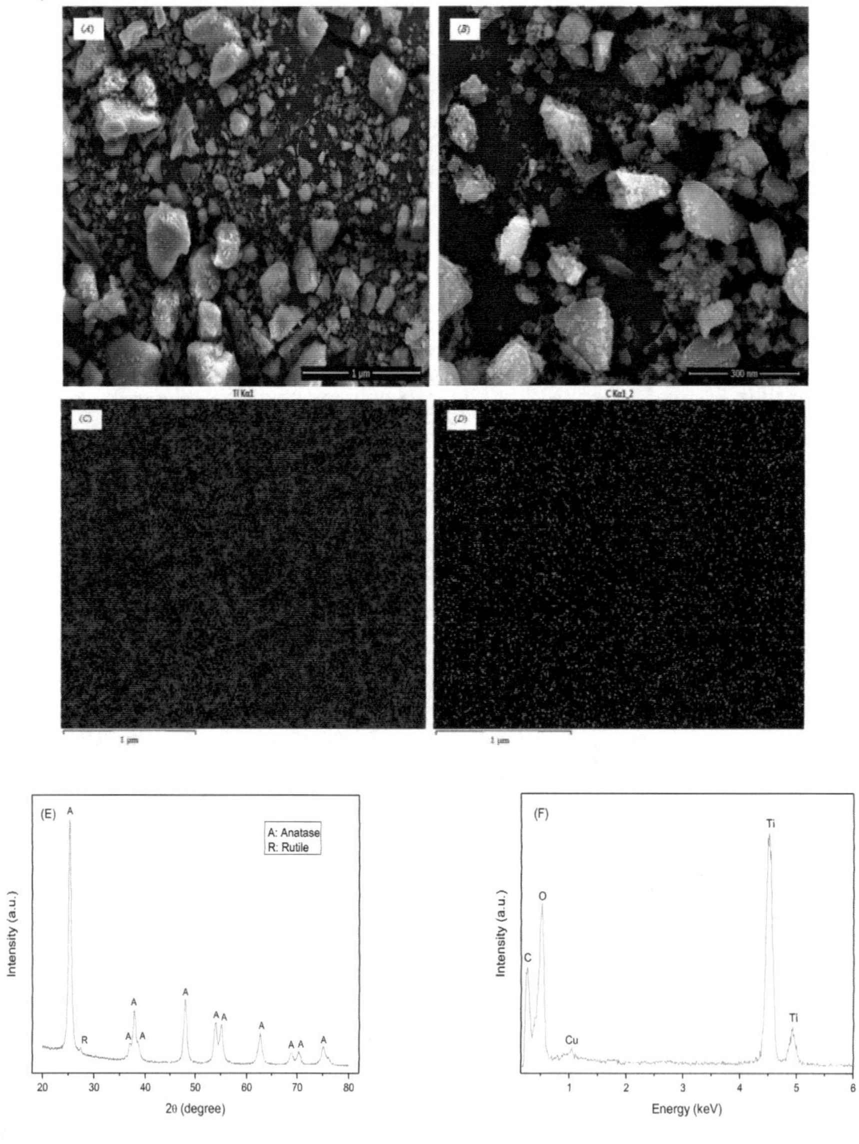

Figure 2. FE-SEM images (*A*,*B*), plane scans (*C*,*D*), XRD pattern (*E*) and EDX spectrum (*F*) of carbon-doped TiO_2.

The XPS survey spectrum in Figure 3*A* shows the C 1*s*, Ti 2*p*, Ti 2*s* and O 1*s* core levels in the TiO_2. The high-resolution XPS spectra (Figures 3*B-E*) show two peaks at 284.1 and 288.1 eV, which indicate the presence of C element; they were assigned to the C-C and C-O bonds, respectively. Additionally, two typical peaks of the Ti-O bond at 458.0 and 463.9 eV were observed, which correspond well with the binding energy of Ti $2p_{3/2}$ and Ti $2p_{1/2}$. The Ti 2*s*-level had a binding energy of 564.6 eV, which superpose the loss structure of O 1*s*. The O 1*s* region was deconvoluted into two peaks located at 530.4 and 529.2 eV, which exhibited the characteristics of O-H and O-Ti bonds. These results proved the successful formation of carbon-doped TiO_2.

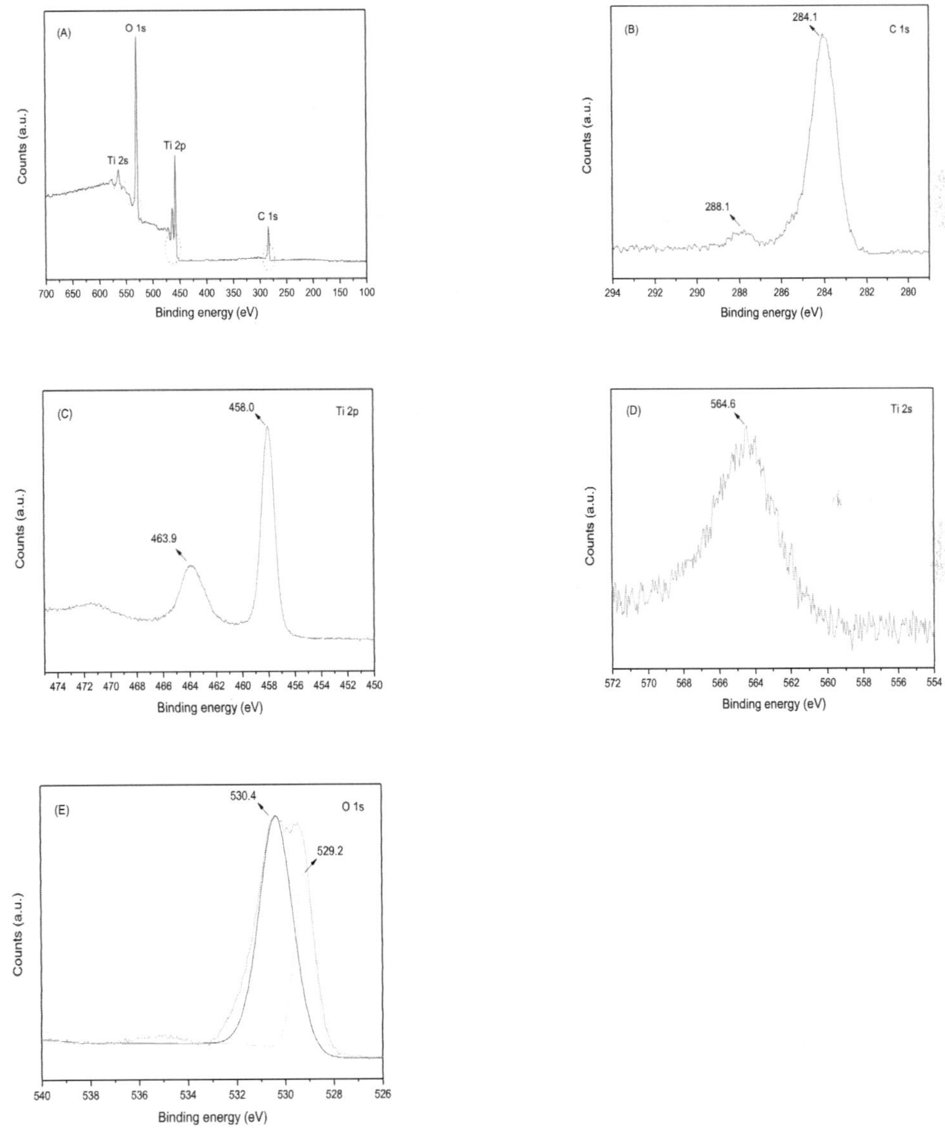

Figure 3. XPS survey spectrum (*A*), and high-resolution XPS spectra of C 1*s* (*B*), Ti 2*p* (*C*), Ti 2*s* (*D*) and O 1*s* (*E*) for carbon-doped TiO_2 as a function of binding energy.

3.2. Photocatalytic evaluation of carbon-doped TiO_2 for formaldehyde

The formaldehyde solution in quartz containers, described in Section 2.4, was placed in the chambers, and the activities of two catalysts were compared by monitoring the formaldehyde degradation as a function of irradiation time (Figure 4). The D_f of the pure TiO_2 was less than 15% in 55 min (an indication of little growth), whereas that of the carbon-doped TiO_2 was 81% after 70 min of irradiation. Carbon-doped TiO_2 is considered to exhibit high catalytic activity due to the generation of hydroxyl radicals and trapped holes.

We hypothesize that the mechanism is as follows: C is incorporated to replace O or Ti in the TiO_2 lattice, which results in a reduction in the bandgap of the TiO_2; this allows photocatalytic reactions to proceed under visible-light irradiation. Herein, the carbon-doped TiO_2 was employed in the degradation of the formaldehyde from wood-based panels with excitation at 410 nm.

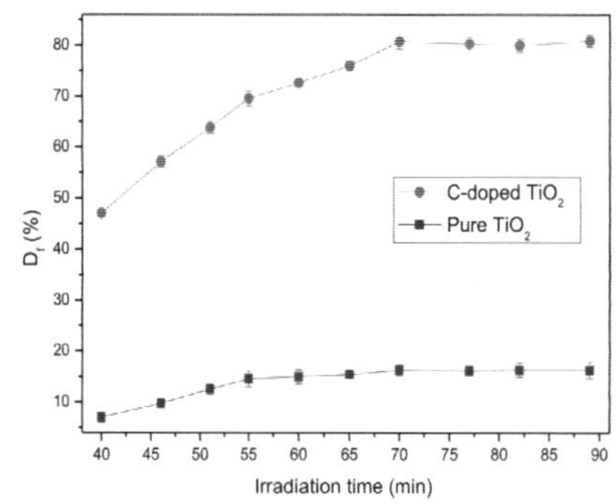

Figure 4. Effect of irradiation time on photocatalytic performance of catalysts under visible light; error bars represent standard deviations ($n = 3$).

The formaldehyde from the panels diffused into the chamber atmosphere. Without heating the chambers, the photodegradation was induced by illumination with the catalysts. The D_f of the representative positive samples over the carbon-doped TiO_2 was determined. As listed in Table 2, the mean D_f obtained are greater than 84%; a maximum of 90% is achieved, which demonstrate efficient removals. This implies that the catalytic application of carbon-doped TiO_2 was reliable, and the treatments with visible-light active catalysts enhanced the control of the formaldehyde at safe levels from the wood-based panels after 70 min of irradiation.

Table 2. Mean degradation rates of formaldehyde from wood-based panels.

No.	A_0 (before photodegradation)	A_t (catalyzed by carbon-doped TiO_2)	Type of panel	D_f(%)
1	0.421	0.062	Multilayer board	85.3
2	0.318	0.049	Medium-density fiberboard	84.6
3	0.325	0.043	Particleboard	86.8
4	0.592	0.058	Plywood	90.2
5	0.587	0.060	Veneer	89.8
6	0.380	0.051	Blockboard	86.6
7	0.602	0.064	Medium-density fiberboard	89.4
8	0.285	0.036	Plywood	87.4
9	0.319	0.050	Blockboard	84.3
10	0.483	0.047	Particleboard	90.3

4. Conclusions

A new approach for hydrothermal synthesis of submicron-sized carbon-doped TiO_2 with enhanced catalytic performance is reported herein. The substitutional carbon was a functional composition in the photocatalytic reactions. The activities of the catalysts in environmental chambers were investigated, and it was revealed that the carbon-doped TiO_2 exhibited higher D_f (81% within 70 min) than the undoped TiO_2 (15% within 55 min). The mean D_f of the representative panels were obtained to be >84%, and therefore the degradation of formaldehyde was promoted under visible light. Hence, carbon-doped TiO_2 is beneficial in the control of formaldehyde emissions from wood-based panels, and it could be utilised in corresponding photocatalytic applications.

Acknowledgements

This work was supported by the Hangzhou Research Project for Agricultural and Social Development of China [2020ZDSJ0110] and the Zhejiang Quality Technology Infrastructure Project of China [20200116].

References

[1] Nielsen G D, Larsen S T, Wolkoff P 2017 *Archives of Toxicology* **91** 35
[2] Miralles P, Chisvert A, Alonso M J, Hernandorena S, Salvador A 2018 *Journal of Chromatography A* **1543** 34
[3] Kwak K, Paek D, Park J 2020 *American Journal of Industrial Medicine* **63** 312
[4] Liu Y, Zhu X, Qin X, Wang W, Hu Y, Yuan D 2020 *Environmental Monitoring and Assessment* **192** 348
[5] Baik S, Lee H J, Kim D W, Min H, Pang C 2019 *ACS Applied Materials & Interfaces* **11** 25674
[6] Kim K W, Kim H J, Park J C 2010 *Journal of Hazardous Materials* **177** 90
[7] Yuan R H, Huang F, Jiang J, Huang J F, Su C 2019 *Toxicological and Environmental Chemistry* **101** 215
[8] Kanan S, Moyet M A, Arthur R B, Patterson H H 2020 *Catalysis Reviews* **62** 1
[9] Li H, Ma A, Zhang D, Gao Y, Dong Y 2020 *RSC Advances* **10** 4681
[10] Xie A J, Zhou X M, Zhou W T, Luo S P, Yao C 2016 *Materials Technology* **32** 265
[11] Zheng J, Liu Z Q, Liu X, Yan X, Li D, Chu W 2011 *Journal of Alloys and Compounds* **509** 3771
[12] Yuan R H, He H H, Huang J F, Su C 2019 *Materials Chemistry and Physics* **235** 121726
[13] Lin Y T, Weng C H, Lin Y H, Shiesh C C, Chen F Y 2013 *Separation and Purification Technology* **116** 114
[14] Putri K, Annisa A, Husain S, Rodiansono R 2020 *Bulletin of Chemical Reaction Engineering and Catalysis* **15** 35
[15] Chen X Y, Sun H Z, Zelekew O A, Zhang J, Guo Y B, Zeng A Q, Kuo D, Lin J G 2020 *Applied Surface Science* **525** 146531

Energy absorption characteristics of honeycomb materials with negative poisson's ratio*

Xiaonan Yan[1], Rong Wang[2,3], Yongxiong Chen[2], Yanhai Cheng[1] and Xiubing Liang[2]

[1]School of Mechatronic Engineering, China University of Mining and Technology, Xuzhou 221116, China;
[2]National Innovation Institute of Defense Technology, Academy of Military Science PLA, Beijing 100071, China

[3]E-mail: wangrongerdos@126.com

Abstract. Using a combination of finite element numerical simulation and experimental verification, the honeycomb structure with different parameters of negative Poisson's ratio is compressed at a uniform speed. The influence of angle parameter θ, size parameter h and thickness parameter t on Poisson's ratio and energy absorption of the whole structure is studied, and the main factors influencing Poisson's ratio and energy absorption are discussed. The results show that the larger concave angle, the smaller size and the larger thickness of the honeycomb cell, the better energy absorption of the overall structure, and the angle is the parameter that has the greatest influence on the Poisson's ratio and energy absorption. Meanwhile, it is demonstrated that the honeycomb hybrid auxetic material has better energy absorption.

1. Introduction

Poisson's ratio refers to the negative number of the ratio of transverse strain to longitudinal strain when a material is longitudinally tensioned or compressed, that is $\nu = -\varepsilon_x / \varepsilon_y$. It was first proposed by French scientist Simeon Denis Poisson (1787—1840). Classical elasticity theory proves that the Poisson's ratio of isotropic materials ranges from -1.0 to 0.5 [1]. Generally, the material will contract laterally when stretched, and expand laterally when compressed [2]. On the contrary, negative Poisson's ratio materials show the characteristics of both transverse and longitudinal contraction or expansion, as shown in Figure 1. Compared with positive Poisson's ratio materials, negative Poisson's ratio materials have some excellent properties, which are specifically manifested in light weight, high damping, energy absorption, elastic modulus and shear modulus [3-4]. It plays an important role in functional materials and is of great significance to the development of transportation, energy and power, military armor, aerospace, and building materials [5-8]. Among them, the hexagonal negative Poisson's ratio honeycomb structure is generally used in the design of marine vibration isolation bases [9]; the three-dimensional multi-cell structure is used in the design of the energy absorption box of the

*Beijing Science and Technology Project (Z191100002719009);
National Key R&D Program of China (2018YFC1902400);
Youth Program of National Natural Science Foundation of China (52005504).

automobile front longitudinal beam [10]; the concave honeycomb structure is used in the structural design of submarine power equipment compartments [11].

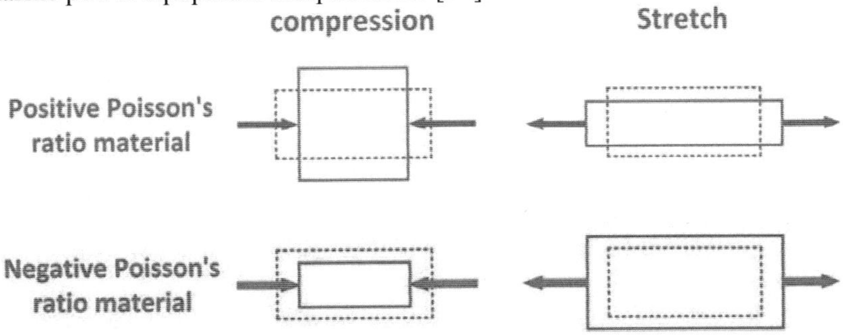

Figure 1. Deformation of positive and negative Poisson's ratio materials.

Due to the excellent mechanical properties of negative Poisson's ratio materials, researchers have conducted extensive research on negative Poisson's ratio metamaterials. In 1987, Lakes first obtained a negative Poisson's ratio foam material through heat treatment of polyurethane foam, and its Poisson's ratio was measured to be -0.7 [12]. Afterwards, researchers have successively proposed concave polygonal structures, rotating rigid body structures, chiral structures, etc., and the cell structure studied has gradually expanded from two-dimensional to three-dimensional [13-15]. At the same time, researchers began to study the cell array problem, and proposed the angle gradient and thickness gradient arrangement. The cell structure of different gradient arrangements is mechanically analyzed, and the advantages and disadvantages of various gradient arrangements are compared [16-17]. Since then, researchers have conducted in-depth studies on the static compression and dynamic impact of materials with negative Poisson's ratio, revealing the principle of force deformation of materials with negative Poisson's ratio [18-19].

The existing research results mainly focus on the design of cell structure, the gradient design of the overall structure, dynamic impact research, etc. These studies have made important contributions to the structural expansion and force analysis of negative Poisson's ratio materials, but the research on parameter optimization and energy absorption of cell structure has just begun. Among them, Chuong N et al. presented an approach for designing material micro-structures by using isogeometric analysis and parameterized level set method. By designing and optimizing two-dimensional and three-dimensional negative Poisson's ratio structures, the effectiveness of the topology optimization algorithm coupled with the reduced basis approach in designing metamaterials is proved [20]. Fábio A. O. F optimized the general helmet by comparing the performance of helmets with different parameters under double impact through numerical simulation. Finally, a helmet lining structure with superior overall performance and capable of withstanding multiple impacts is obtained [21]. These studies have made the material have better energy absorption and structural performance. But they have not studied the changes of Poisson's ratio and energy absorption under the change of geometric parameters, and the main parameters that affect energy absorption and Poisson's ratio.

As the core of negative Poisson's ratio metamaterials, the cell structure of negative Poisson's ratio directly determines the overall performance, so the optimization of cell structure parameters is very important. The honeycomb structure is widely used in structural applications due to its special strength. At the same time, considering the regularity of structural cells and the ease of production, this paper chooses the concave honeycomb structure as the research object. Through the combination of finite element numerical simulation and experiment, a concave honeycomb model with different parameters is established. In-depth study of the effect of the concave angle, thickness, and size of the cell geometric parameters on the Poisson's ratio value of the structure, and compare the energy absorption according to the compressive stress-strain curve. The research results show that the larger concave angle, the smaller size, and the larger thickness of the cell, the better energy absorption of the overall structure, and the angle is the parameter that has the greatest influence on the Poisson's ratio and energy absorption. In addition, a metal-ceramic hybrid auxetic structure is proposed, and it is confirmed that the structure has better compression resistance, which provides a certain sense of guidance for the design and application of negative Poisson's ratio materials.

2. Models and methods

2.1. Specimen fabrication

Figure 2 shows the cell structure of the concave honeycomb. In the figure: h and l represent the length of the horizontal cell wall and the length of the inclined cell wall respectively, θ is the concave angle, t is the thickness of the cell wall, and $h/l=2$. When the concave honeycomb structure is longitudinally compressed, the diagonal ribs of the honeycomb unit will deform and gradually move from the diagonal direction to the horizontal direction, so that the material exhibits a negative Poisson's ratio effect. In fact, the main reason for the negative Poisson's ratio of the honeycomb material is the axial compression and hinged rotation, which causes the diagonal ribs to bend inward and cause the structure to contract both laterally and longitudinally.

Figure 3 shows the sample with a concave angle of -30°, printed by a flash-cast dreamer 3D printer with a precision of 0.1mm produced in Zhejiang. The printing parameters are shown in Table 1. The printing material is produced by Dake Intelligent and the hardness is 95A TPU material with a diameter of 1.75mm, 100% filling and printing is used at 200° to maximize the hardness. The printed model is arrayed with 3 cells in the high direction and 7 cells in the width direction, with a uniform depth of 20mm. When studying the influence of the concave angle on the structural performance, h=11.5mm and t=1mm are uniformly adopted, and the angle θ is respectively 30°, 20°, 10°, 0°, -10°, -20°, -30° seven types. A positive value represents a conventional honeycomb structure, and a negative value represents a concave honeycomb structure. All samples are printed by the same 3D printer to ensure the consistency of the printed sample quality.

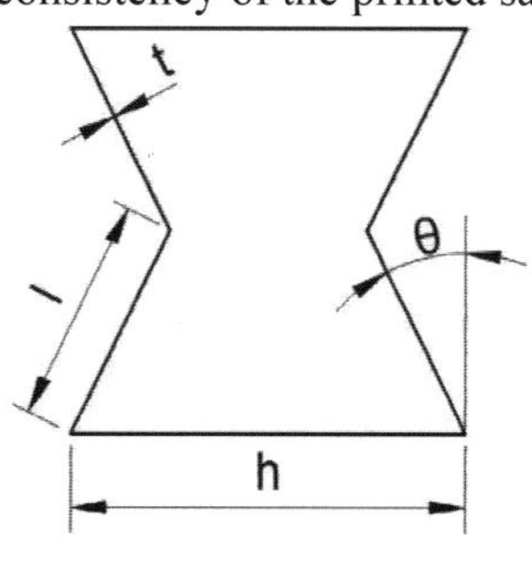

Figure 2. Cell structure of concave honeycomb.

Figure 3. Concave Angle -30°sample.

Table 1. 3D printer printing parameters.

parameter name	printing speed/(mm·s^{-1})	Printing thickness/ mm	Nozzle temperature/℃	Hot bed temperature/℃	Withdrawal speed/(mm·s^{-1})
Parameter index	45	0.5	220	70	80

To study the compressive performance of materials, a universal testing machine is usually used to do a compression test to obtain a stress-strain curve. The area enclosed by the stress-strain curve and the coordinate axis represents the energy absorbed by the material [22]. The calculation method is calculated by formula (1):

$$\omega = \int_{0}^{\varepsilon_0} \sigma \mathrm{d}\varepsilon \quad (1)$$

In the formula: ω is the energy absorbed by the material; σ is the compressive stress; ε is the compressive strain; ε_0 is the upper limit of the compressive strain.

2.2. Finite element modeling

In this paper, ABAQUS finite element simulation software is used for numerical simulation. The simulation diagram is shown in Figure 4. The middle of the model is a concave honeycomb structure, and the upper and lower sides are two rigid plates. The simulation uses ABAQUS's display dynamics analysis method to compress the structure slowly and uniformly.

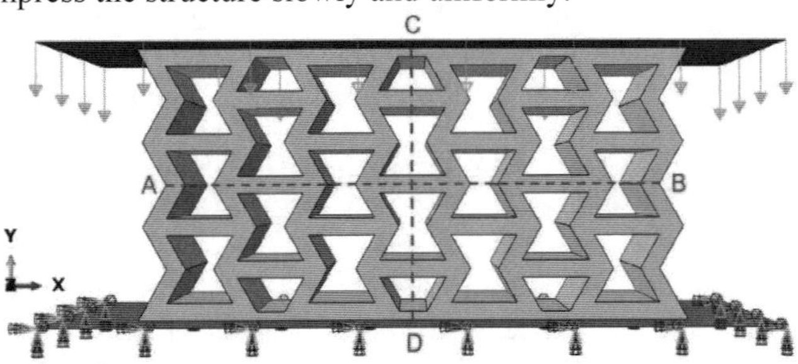

Figure 4. Schematic diagram of compression simulation of honeycomb structure.

The model material is TPU material, which is assumed to be an ideal elastoplastic model. The material parameters are measured by the tensile experiment, and the tensile stress-strain curve is shown in Figure 5. According to the tensile test, its material parameters are: Young's modulus E=207MPa, Poisson's ratio u=0.25, yield stress σ_y=0.154MPa, and density ρ= 1100kg/m^3. Both the compression end rigid body and the fixed end rigid body adopt discrete rigid bodies, and the upper and lower rigid bodies are bound to the structure. The upper rigid body is pressed down slowly at a uniform speed, and is restricted in the other directions. The compression end rigid body and the fixed end rigid body are both discrete rigid bodies, and the fixed end rigid body is completely fixed (U1=U2=U3=UR1=UR2=UR3=0), and the upper and lower rigid bodies are bound to the structure. The displacement curve of the upper rigid body is y=0.032t, and it is restricted in the other directions. In order to prevent the materials from penetrating each other after compression [23], the whole structure adopts universal contact. The coefficient of contact friction is 0.2, and they are in hard contact with each other. In order to improve the quality of the mesh and the convergence of the analysis, the element shape is a hexahedron, and the meshing type is C3D8R. The average size of the unit is 1mm, and the structure is divided into 21200 units. The result of the structural grid division is shown in Figure 6. After compression deformation, the transverse strain ε_{11} and the longitudinal strain ε_{22} are obtained by measuring the deformation values of structure width AB and height CD, and then dividing the original length AB and CD. Finally, the Poisson's ratio of the compression model is obtained by Formula $\nu = -\varepsilon_{11}/\varepsilon_{22}$.

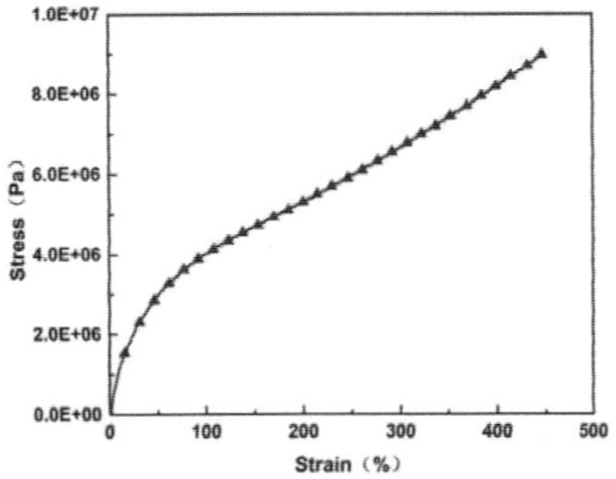

Figure 5. Stress-strain curve of TPU material.

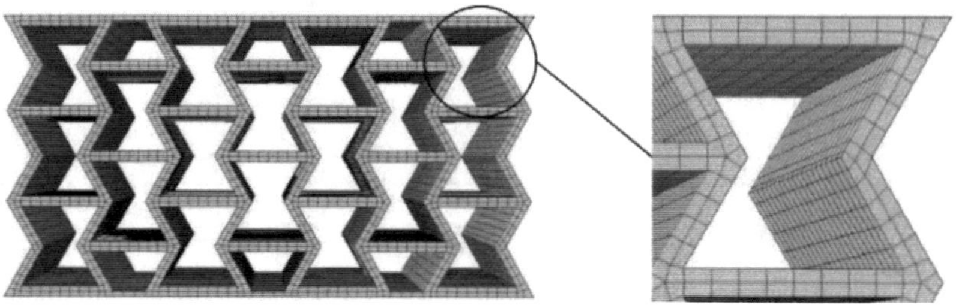

Figure 6. Grid division of honeycomb structure.

2.3. Experimental program

In this experiment, the MTS-20G microcomputer-controlled electronic universal testing machine is used, and the maximum test force is 20KN. In the experiment, the standard compression speed is 0.001% for compression, which is exactly the same as the compression speed in the simulation. The longitudinal stress and strain values are given by the computer during the compression process.

Figure 7 shows the test and simulation diagram of a sample with a concave angle of -30° and a compression of -10mm. It can be seen from the figure that the deformation of the structure in the experiment and simulation is similar. When the structure is compressed longitudinally, there is a concave phenomenon in the transverse direction, and the concave in the middle of the structure is the most obvious. Figure 8 shows the stress-strain curve of the specimens compression process with a concave angle of -30° and 30°. From the stress-strain curve, it can be seen that there is a certain error deviation between the test and the simulation, but the overall trend of the two curves is the same. The peak stress reached by compression is also very similar, indicating that the simulation data is accurate.

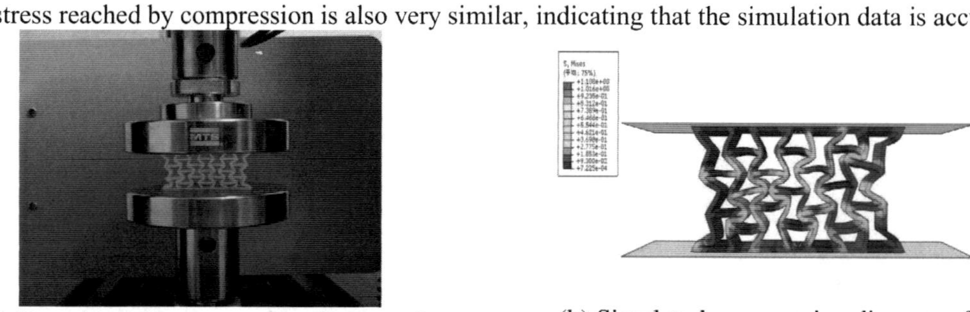

(a) Experimental compression diagram of specimen with concave angle of -30°

(b) Simulated compression diagram of specimen with concave angle of -30°

Figure 7. Concave angle -30°deformation of the sample after compression.

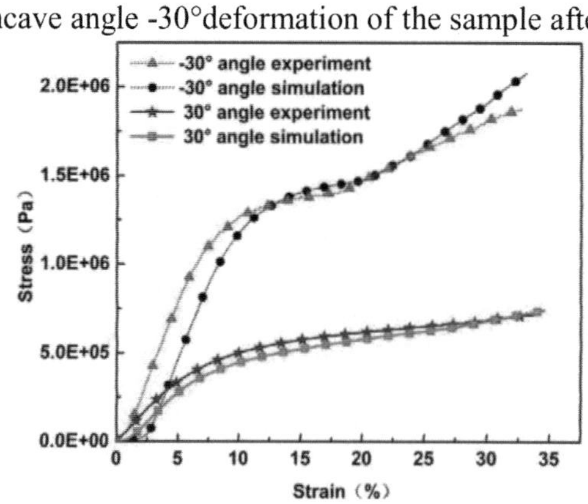

Figure 8. Stress-strain curves of specimens with concave angles of -30° and 30°.

3. Results and discussion

3.1. *Effect of angle parameter*

After the experimental compression of 10mm, the seven-angle Poisson's ratio values are shown in Figure 9. It can be seen from the figure that the Poisson's ratio values are symmetrical, reaching the maximum and minimum values at 10°and -10°. The reason why the maximum value is reached at these two points is because the concave angle is small. During compression, the angle change is large, and the horizontal and vertical strain will also increase, so the Poisson's ratio reaches the maximum value. Figure 10 shows the stress-strain curves of seven angle experiments. It can be seen that the negative curve is significantly higher than the positive curve. The stress-strain curve of the concave honeycomb structure is significantly higher than that of the ordinary honeycomb structure, which means that the concave structure can absorb more energy, so the concave structure has an excellent energy absorption. Comparing the Poisson's ratio and energy absorption diagrams of the seven angles, it can be seen that as the angle becomes gradually concave, the value of Poisson's ratio first decreases and then increases. But at the same time the height of the stress-strain curve is rising,and the area enclosed is also larger. This shows that the greater concave angle, the greater energy absorbed by the structure. But it does not correspond to the smallest Poisson's ratio, indicating that the smaller the Poisson's ratio, the best energy absorption.

It can be seen from the stress-strain curve that the 0° and -10° structures have obvious softening phenomenon while the others are not obvious. The greater the absolute value of the angle, the less obvious the softening phenomenon. This is because the angle is small and the structure is not easy to deform during compression, so the stress value rises faster. However, when the structure deforms along the concave angle, the stress value will decrease and the softening phenomenon will occur. The structure with a large absolute value of angle is easy to be deformed along the concave angle during compression, so there is no softening phenomenon and the stress value rises gently. This also indicates that when the structure with small concave angle is subjected to angular compression, the anti-compression ability of the structure will decrease, which is far less than that of the structure with large concave angle.

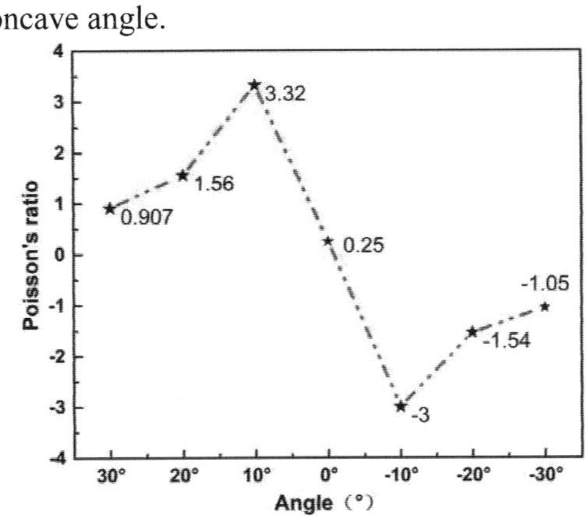

Figure 9. Poisson's ratio of seven angles.

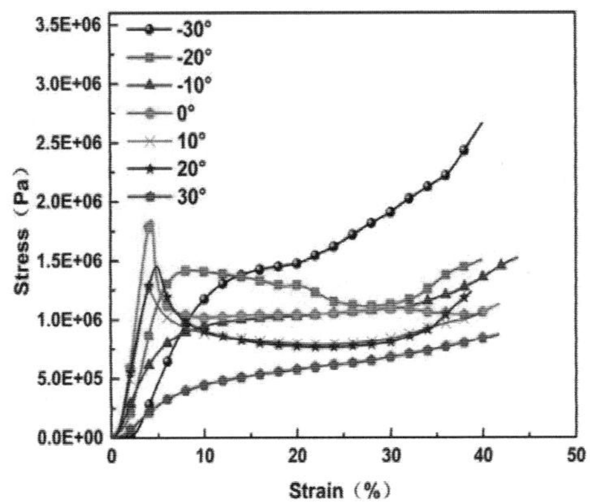

Figure 10. Stress-strain curves of seven angles.

3.2. *Effect of size parameter*

While studying the influence of angle parameters on the structure, it also demonstrated the correctness of the simulation. Therefore, four structures with cell length h of 2mm, 4mm, 8mm and 16mm were selected for the simulation compression test. At this time θ=-30°, t=1mm remains unchanged, the simulation result is shown in Figure 11. With the gradual increase of the cell length h, the Poisson's ratio is continuously reduced, and the peak stress is also reduced. The value of Poisson's ratio almost coincides with the value of stress, which shows that the value of Poisson's ratio and the peak stress are synchronized with the cell size. It can be seen from the figure that the stress value is the largest when the cell length is 2mm, so it shows that the smaller cell size, the better energy absorption of the structure. But at this time the Poisson's ratio of the structure is the largest. At this time, it also shows that the smaller Poisson's ratio of the structure, the energy absorption is not the maximum.

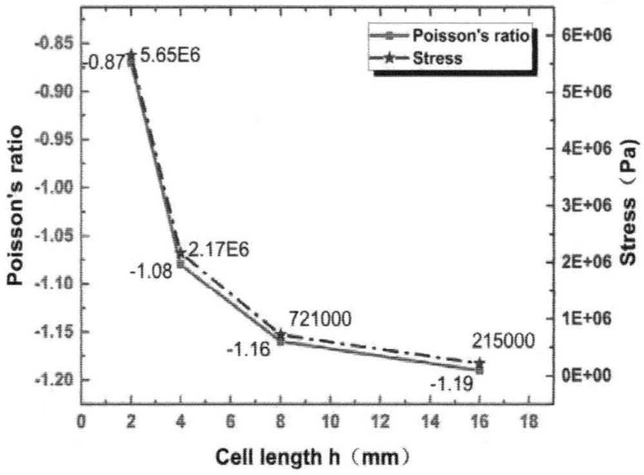

Figure 11. Poisson's ratio and stress after compression of different cell lengths.

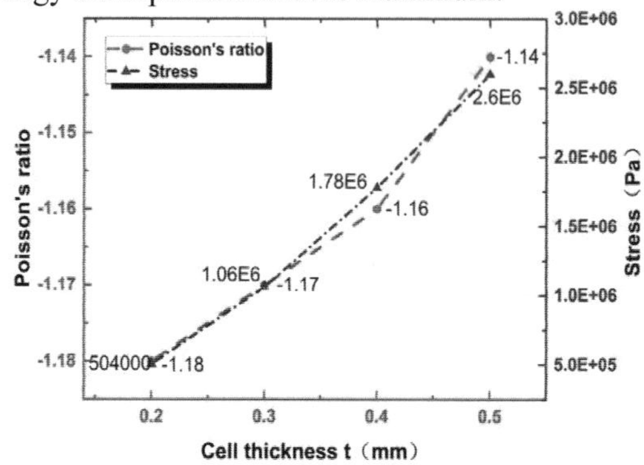

Figure 12. Poisson's ratio and stress after compression of different cell thicknesses.

3.3. Effect of thickness parameter

Four structures with thicknesses of 0.2mm, 0.3mm, 0.4mm and 0.5mm were selected for compression simulation, while θ=-30°, h=8mm remained unchanged, and the simulation results are shown in Figure 12. It can be seen from the figure that as the cell thickness increases, the Poisson's ratio of the structure is increasing, and the stress value is also increasing. This conclusion is the same as that of Yang D U's research. The value of Poisson's ratio and stress value also have the same trend as the simulation result of length, and both are synchronous and almost coincide. This shows that the greater thickness of the cell, the greater Poisson's ratio of the structure. Moreover, the stress value at this time is also large, and the energy absorption is also the best. This also shows that for a negative Poisson's ratio structure, it is not that the smaller Poisson's ratio, the better energy absorption.

4. Discussion and extension

Each parameter affects the performance of the overall structure, but the decision of which of the three parameters needs further discussion and analysis. Only knowing the decision parameters can easily change the parameters to improve the performance of the overall structure.

4.1. The main factors affecting Poisson's ratio

Figure 13 shows the influence of three parameters on Poisson's ratio. From the figure, it can be seen that it is the concave angle that determines the negative Poisson's ratio of the structure. Only when the concave angle is negative can there be negative Poisson's ratio. Secondly, it can be observed that the angle has the greatest influence on Poisson's ratio, which is much larger than the other two parameters and plays a leading role. Moreover, the Poisson's ratio has a linear relationship with the size h and thickness t, but the angle is not. Although the 0° angle is closest to the zero degree line, the angles on both sides are symmetrically distributed. Moreover, the closer the negative concave angle is to 0°, the greater the Poisson's ratio, so it is not that the larger concave angle, the greater Poisson's ratio.

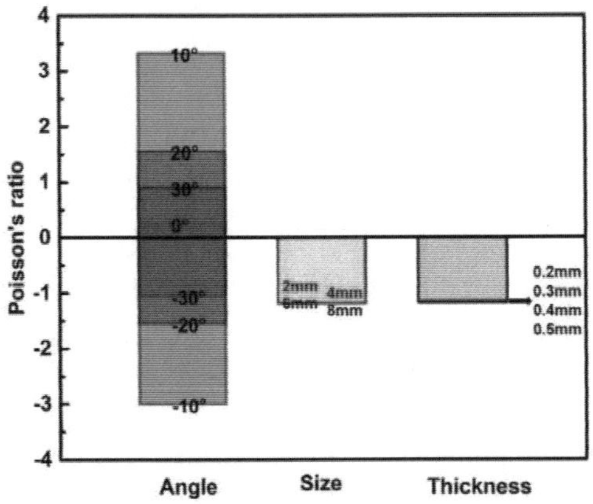

Figure 13. Diagram of factors affecting Poisson's ratio.

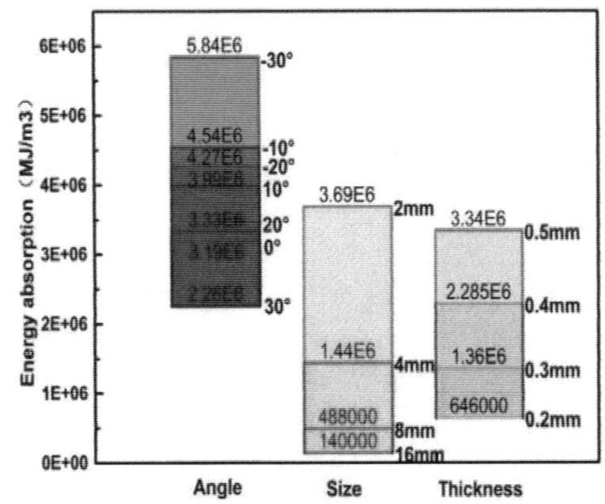

Figure 14. Diagram of factors affecting energy absorption.

4.2. The main factors affecting energy absorption

Figure 14 is a graph of the influence of three parameters on energy absorption. It can be seen from the figure that thickness has the smallest influence on energy absorption, while angle and size have a relatively large influence on energy absorption. However, there are obvious differences between the two parameters. The size data is obviously lower than the angle data, which shows that the influence of the angle is much greater than the other two parameters. A slight modification of the angle can greatly improve the energy absorption of the overall structure. It can also be seen from the figure that when the upper value of the three parameters is taken, namely θ=-30°, h=2mm, and t =0.5mm, the energy absorption of the structure can reach the best.

4.3. Numerical simulation of concave honeycomb hybrid auxetic material

Ceramic materials have good mechanical properties such as low density, high strength and high hardness, and are good materials for impact resistance [24-25]. The metal-ceramic hybrid auxetic material formed by combining the metal honeycomb structure and the ceramic can not only exert the advantages of the high strength of the ceramic material, but also exert the concave characteristic of the honeycomb structure. The external metal concave structure can well wrap the broken ceramics, making the hybrid auxetic material an ideal material with light weight and high energy absorption.

Based on the above research conclusions, the energy absorption of negative Poisson's ratio honeycomb material, metal-ceramic hybrid auxetic material and metal solid material was compared through numerical simulation, and the best energy absorption material was finally obtained. The three simulated structural materials are shown in Figure 15. The metals of the three structural materials are all alumina materials, and the ceramic materials are SiC. The parameter setting in the simulation is the same as the above, with a slow and uniform pressure of 5mm.

(a)Honeycomb material (b) Hybrid auxetic material (c) Solid material

Figure 15. Three simulated structural materials.

The compression results of the three structural materials are shown in Figure 16. It can be seen that the three materials have undergone certain deformation. The concave phenomenon of honeycomb structure is obvious. In the hybrid auxetic material structure, the ceramic is broken and the outer metal is recessed. The solid structure is simply a compression deformation. Figure 17 shows the stress-strain curves of the three materials during compression. It can be clearly seen from the figure that the hybrid auxetic structure has the highest stress value, followed by the solid structure. By comparison, it can be found that the hybrid auxetic structure absorbs the most energy, which also shows that the hybrid auxetic structure has the best compressive performance. This is because the ceramic has high strength and the metal honeycomb structure can well wrap the broken ceramics, so the hybrid auxetic structure has the best mechanical performance.

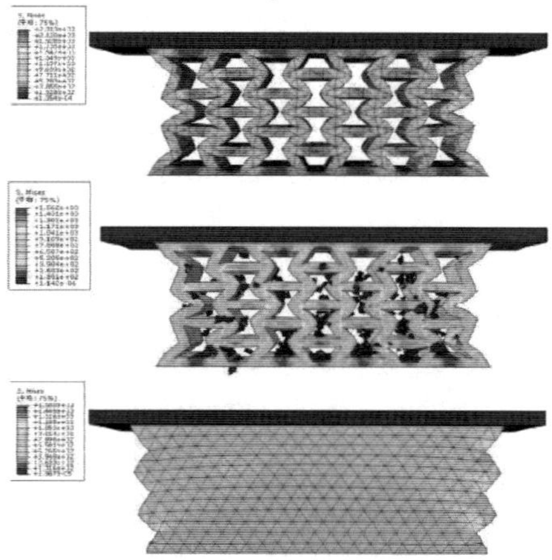

Figure 16. Compression results of three materials.

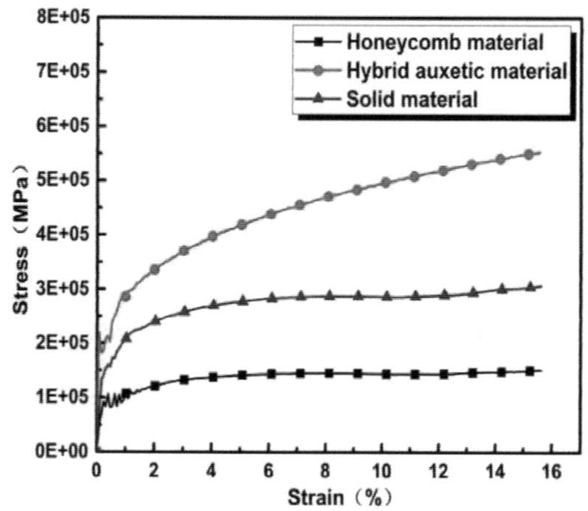

Figure 17. Compression stress-strain curves of three materials.

5. Conclusions

In this paper, the different angle parameters θ, size parameter h and thickness parameter t of the concave honeycomb structure are simulated and experimentally studied respectively. The effect of the three parameters on the overall structure is comparatively studied, and the main factors affecting energy absorption and Poisson's ratio are discussed. Get the following conclusions:

(1) The larger concave angle of the concave honeycomb structure, the smaller size and the thicker thickness, the better energy absorption of the honeycomb structure. It can absorb more external pressure and improve the pressure and energy absorption of the material. In addition, it is also pointed out that no matter which parameter changes, it cannot be explained that the smaller the Poisson's ratio, the best energy absorption. There is no direct relationship between the two.

(2) It explains why the energy absorption is the best under the three parameters discussed in the article. This is because the concave angle is large and it is not easy to compress and deform. The small size structure is dense and compressive, and the thickness of the structure is large, so the energy absorption is the best under these three parameters.

(3) Through comparative studies, it is found that angle is the parameter that has the greatest influence on energy absorption and Poisson's ratio. Therefore, to change the performance of the overall structure, the adjustment of the angle should be considered first.

(4)The honeycomb concave structure composited with ceramic has better compressive performance, because the ceramic has high strength and the honeycomb structure can well wrap the broken ceramics.

References

[1] Sokolnikoff I S 1983 Mathematical theory of elasticity *Malabar:Krieger Publishing Company*
[2] Yang D U, Lee S and Huang F Y 2003 Geometric ffects on micropolar elastic honeycomb

structure with negative Poisson's ratio using the finite element method *Finite Elements in Analysis and Design* **39(3)** 187-205
[3] Grujicic M, Galgalikar R, Snipes J S, et al. 2013 Multi-physics modeling of the fabrication and dynamic performance of all-metal auxetic-hexagonal sandwich-structures **51** 113-130
[4] Tomasz S, Hubert J, Bogdan T M, et al. 2014 Computational analysis of sandwich - structured composites with an auxetic phase **251(2)** 354-366
[5] Hui W, Yuxuan Z, Wanqing L and Qinghua Q 2020 A novel two-dimensional mechanical metamaterial with negative Poisson's ratio *Computational Materials Science* **171** 0927-0256
[6] Alderson A and Alderson K L 2007 Auxetic Materials *Proceedings of the Institution of Mechanical Engineers Part G Journal of Aerospace Engineering* **221(4)** 565-575
[7] Heo H, JU J and Kim D M 2013 Compliant Cellular Structures: Application to a Passive Morphing Airfoil *Composite Structures* **106(12)** 560-569
[8] Jacobs S, Coconnier C, DiMaio D, et al. 2012 Deployable auxetic shape memory alloy cellular antenna demonstrator: design, manufacturing and modal testing **21(7)**
[9] Binghong W, Xiangwen Z and Deqing Y 2018 Analysis of real ship application of negative Poisson's ratio metamaterial vibration isolation base *Ship Engineering* **40(02)** 56-62
[10] Xing Y, Ye Y, Wei Z and Wenbin H 2017 Optimal design of automobile energy-absorbing box based on three-dimensional multi-cell structure *Journal of Dalian University of Technology* **57(04)** 331-336
[11] Lifu X and Deqing Y 2019 Analysis of submarine vibration and sound radiation performance of metamaterials with negative Poisson's ratio *Journal of Vibration Engineering* **32(06)** 956-965
[12] Lakes R 1987 Foam structures with a negative Poisson's ratio *Science* **235(4792)** 1038-1040
[13] Miller W, Hook P B, Smith C W, Wang X and Evans K E 2009 The manufacture and characterisation of a novel, low modulus, negative Poisson's ratio composite *Compos. Sci. Technol.* **69 (5)** 651-655
[14] Babaee S, Shim J, Weaver J C, Chen E R, Patel N and Bertoldi K 2013 3D Soft metamaterials with negative Poisson's ratio *Adv. Mate.* **25 (36)** 5044-5049
[15] Jingjun Y, Yan X and Xu P 2018 Research progress of negative Poisson's ratio metamaterials *Chinese Journal of Mechanical Engineering* **54(13)** 1-14
[16] Shufrin I, Pasternak E and Dyskin A V 2015 Negative Poisson's ratio in hollow sphere materials *Int. J. Solids Struct.* **54** 192-214
[17] Guansheng Y and Zhaonan Y 2017 Analysis of impact dynamics performance of gradient negative Poisson's ratio honeycomb material *Journal of Dynamics and Control* **15(01)** 52-58
[18] Huilong H, Xinchun Z and Peng W 2019 Dynamic response and energy absorption characteristics of honeycomb materials with negative Poisson's ratio *Explosion and Shock* **39(01)** 47-57
[19] Milton G W 2013 Complete characterization of the macroscopic deformations of periodic unimode metamaterials of rigid bars and pivots *Mech.Phys.Solids.* **61(7)** 1543-1560
[20] Chuong N, Xiaoying Z, Ludovic C, et al. 2020 Three-dimensional topology optimization of auxetic metamaterial using isogeometric analysis and model order reduction **371**
[21] Fábio A. O. F, Ricardo J. Alves de S, Mariusz P, et al. 2019 Helmet Design Based on the Optimization of Biocomposite Energy-Absorbing Liners under Multi-Impact Loading **9(4)**
[22] Gibson L J and Ashby M F 1997 Cellular solids: Structure and properties *2nd ed. Cambridge:Cambridge University Press* 87-148
[23] Xiaolin D 2016 In-plane impact dynamics analysis of a honeycomb structure with a layered ladder variable negative Poisson's ratio *Machine Design and Manufacturing* **(04)** 219-223
[24] Kaufmann C, Cronin D, Worswick M, et al. 2001 Influence of Material Properties on the Ballistic Performance of Ceramics for Personal Body Armour **10(1)**
[25] Ullah I, Brandt M and Feih S 2016 Failure and energy absorption characteristics of advanced 3D truss core structures **92** 937-948

Preparation and characterization of rabbit hair keratin hydrogel

NaiWei Ge, Yi Zhang[1], Hao Zhang, Ruoying Zhu and Xueping Shi

School of Textile Science and Engineering, Tiangong University, Tianjin 300380, China

[1]Email: tianjinzhangyi@126.com

Abstract. Protein-based hydrogel with the characters of environmentally friendly, resource saving and excellent biocompatibility has become a hot topic in the field polymer materials. In this paper, the rabbit hair keratin was extracted from waste rabbit hair ang then prepared into hydrogel by means of heating-cooling. The conclusion is that the molecular weight of rabbit hair keratin measured by SDS-PAGE concentrated in 40~60kDa and rabbit hair keratin hydrogel in 11~22 kDa. The forming conditions such as concentration and temperature of preparing hydrogel were experimented by using variable-controlling approach, the formation concentration of keratin should be over 125mg/ml and the temperature over 55℃. Various characterization techniques were used to determine the physical and chemical property of the hydrogel. As proved by scanning electron microscopy (SEM), covalent bonds and three-dimensional porous structure with many interconnected pores had been successfully formed. They were also characterized by thermogravimetric analysis (TG) and swelling test, etc. The consequence showed that the equilibrium moisture content can reach up to 20.21% and the swelling rate as high as 9.55%, which were inversely proportional to the concentration of keratin. Furthermore, the hydrogel had good thermal stability, the initial decomposition temperature is around 204℃. It is indicating that the excellent properties of rabbit hair keratin hydrogel can be further studied.

1. Introduction

Hydrogel is a kind of super-absorbent materials formed by polymer monomers crosslinking by physical or chemical methods. The three-dimensional porous structure, that acts as a skeleton, inside hydrogel can lock in moisture by surface tension. So, hydrogel is a polymer material with strong water absorption and good water retention property. There are numerous hydrophilic groups attached to the network structure to remain hydrophilicity [1]. During absorbing water, the polymer chains form a gel network by crosslinking, which makes hydrogel not to dissolve. That is, hydrogel can absorb water that is hundred times of itself and only volume expands, not dissolves [2]. When the hydrogel full swelling, the softness of it is very similar to the human organic tissues`. Hydrogel also has high permeability to sugar, vitamin, oxygen and other nutrients [3]. Besides, some hydrogels have low hardness and high elasticity, some show good biocompatibility and biodegradability. Based on these characters, hydrogel material can be widely used as a raw material in the area of surgical dressing [4], Human tissue scaffold [5], Agricultural drug release [6], Dye processing [7], metal ion absorption [8] and other fields. However, most widely used hydrogel materials cannot meet the requirements of environment-friendly and resource-saving currently, whose degradation products are toxic and harmful to humans and biocompatibility is poor [9,10]. These limit the application scope of the materials severely. Finding a natural bio-based hydrogel material becomes a hot topic.

Keratin is a kind of structural proteins, which extensively exists in nature. It is usually divided into hard keratin and soft keratin. The hard keratin with a sulphur content among 4%~8% is founded in hair, feathers, hooves, shells claws and horns. The sulphur content of soft keratin is about 2%, which exists in the epidermal skin. Rabbit hair belongs to hard keratin, it has a mass of cystines, forms a disulphide bond by crosslinking, and plays an important role in the physical and chemical properties of keratin. The rabbit hair textile industry in China produces hundreds of tons of rabbit hair waste every year [11]. These scraps are rich in proteins in which the content of keratin is above 90% [12]. If people recycle the rabbit hair, extract the keratin and prepare hydrogel using the materials, it can not only reduce the problem of resource-wasting, but also solve the shortcoming such as poor biocompatibility of hydrogel materials.

This study choose keratin extracted from waste rabbit hair as raw material, method of heat-cooling to prepare the keratin hydrogel materials. Then test the property of swelling, thermal stability and any others, and aim to seek for a way to product a natural non - polluting protein - based hydrogel.

2. Material and methodology

2.1. Materials
Rabbit hair (Shandong) used to extract keratin. Urea, NaOH, L-cysteine, Ethanol, KBr and Polyethylene glycol 20000 were purchased from Tianjin kermel chemical reagent co. LTD

2.2. Production of keratin hydrogel
Weight the keratin powder for 1.5g, 1.75g and 2.0g respectively, add 10ml distilled water and 0.03g L-cysteine, put the breaker into a thermostatic magnetic mixer, heat for 30 min, then cool down at room temperature. It produces three different concentrations hydrogels as 125mg/ml, 150mg/ml and 175mg/ml.

For another set of experiments, keep the mass fraction consistent at 15%, change the temperature in 55℃、65℃、75℃, repeat the above steps and then produce the hydrogels.

2.3. Analysis and characterization

2.3.1. SDS-PAGE. The molecular weight distribution of the rabbit hair keratin was determined by sodium dodecyl sulphate polyacrylamide gel electrophoresis (SDS-PAGE). First,15% separation gel and 5% concentrated gel were prepared on the basis of the proportion in Table 1, injected into a mould, placed for a period of time and waited the solution becoming gelatinous. Next, added 10μl of samples into the sample tank, set the constant voltage of 80V until the sample passed the concentrated gel, next adjust the voltage to 120V, till the text finished. Next, took out the adhesive block and dyed in the dye solution for 3hours. Then, decolorized in the decolorizing solution until the bands of the keratin samples can be observed clearly.

2.3.2. SEM. The surface morphology of keratin gel was observed by scanning electron microscope (SEM). Put the samples of keratin powder and hydrogel into the sample chamber, adjusted the contrast ratio, brightness and focal length until the image clear, and then took the pictures.

2.3.3. The swelling ratio. Gravimetric method is used to calculate the swelling ratio of the keratin hydrogel at room temperature. Select distilled water and PBS buffer solution for swelling experiment, respectively. Weigh 8g NaCl, 0.2g KCl, 3.63g Na_2HPO_4 and 0.24g KH_2PO_4 in 1L distilled water as buffer solution (pH= 7.2). The initial weight of the two pieces of the fully dried hydrogel is both denoted as M_0, one is soaked in 20ml distilled water and another in PBS buffer solution. Weigh once an hour until the weight stable. The first one is denoted as M_1, another as M_2. The swelling ratio in distilled water (R) of keratin hydrogel was calculated using the following equation (1):

$$R=\frac{M_1\text{-}M_0}{M_0}\times 100\% \quad (1)$$

Where M_0 and M_1 denote the weight of the freeze-dried hydrogel before and after being submerged in the distilled water, respectively.

And the swelling ratio in PBS buffer solution (SR) of keratin hydrogel was calculated using the following equation (2):

$$SR=\frac{M_2\text{-}M_0}{M_0}\times 100\% \quad (2)$$

Where M_2 denotes the weight of the freeze-dried hydrogel after being submerged in the PBS buffer solution.

Table 1. The proportions of 15%separation gel and 5%concentrated gel

Solution	15% separation gel	5% concentrated gel
Distilled water	1.1ml	2.1ml
30% Acrylamide	2.5ml	0.5ml
10% SDS	50μl	30μl
1.5mol/L Tris-HCl buffer solution	1.3ml	*
1mol/L Tris-HCl buffer solution	*	0.38ml
0.5mol/L Tris-HCl buffer solution	*	*
Tetramethyl ethylenediamine (TEMED)	2μl	3μl
10%Ammonium persulfate (AP)	50μl	30μl

2.3.4. TGA (thermogravimetric) analysis of keratin hydrogel. The thermal property of the samples was measured by TGA in the temperature range of 30 and 600°C at a heating rate of 10K/min in N_2 by using a NETZSCH1500(Shanghai, China).

3. Results and discussion

3.1. SDS-PAGE

Figure 1 shows the molecular weight of keratin powder and hydrogel extracted from rabbit hair. Keratin is a kind of natural polymer protein, but the relative molecular weight of keratin hydrogel extracted from rabbit hair is mainly concentrated in 11 ~ 22KDa by SDS-PAGE gel electrophoresis, which indicates that most of the keratin with high relative molecular weight has been decomposed.

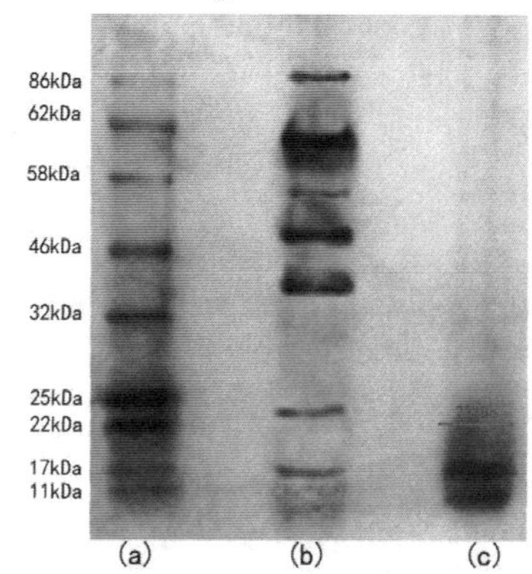

Figure 1. SDS-PAGE patterns: (a) molecular weight standard; (b) keratin powder; (c) keratin hydrogel.

IOP Conf. Series: Materials Science and Engineering 1040 (2021) 012003 doi:10.1088/1757-899X/1040/1/012003

3.2. SEM

Figure 2 shows the different microstructures between the keratin powder with the keratin hydrogel. (a) is the keratin powder which is full of irregular and lamellar structures, and (b)is the keratin hydrogel. It has an obvious non-uniformed porous structure and a multi-layered cross section. According to the two pictures, it can be seen that in the process of the keratin hydrogel preparation, the chemical bonds in keratin molecules are broken and intermolecular forces are generated, forming a network structure. There is good connectivity of the pores, so it has good adsorption, permeability, water retention, etc.

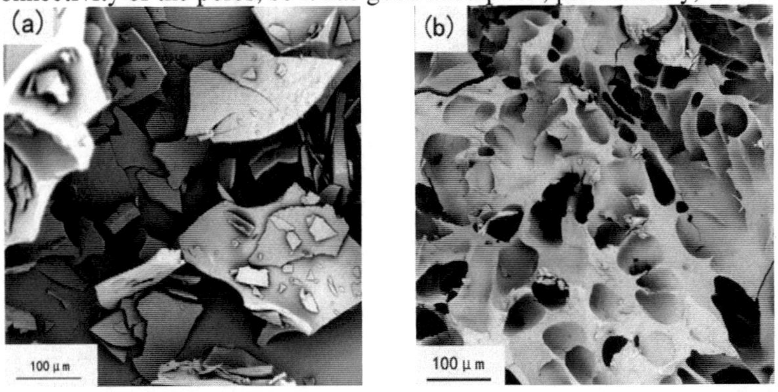

Figure 2. the microstructures of the keratin powder and hydrogel.

The molecular chain structure of rabbit hair keratin is mainly α-helix. The basic feature of α-helix keratin is that it contains a large numerous of cysteine residues, some react to form disulphide bonds, and others are free. There are also hydrophobic interactions and hydrogen bonds in it. Figure 3 shows the inter-and intra-molecular bonding in keratin.

When we add cysteine into the solution of rabbit keratin and raise the temperature, the inter-and intra-molecular bonding break, and then some bonds self-assemble. As a reducing agent, the sulfhydryl groups(-SH) can break the disulphide bond (-S-S-), then form a stable β-sheet conformation by physical crosslinking. Figure 4 shows the scheme of proposed self-assembly of rabbit keratin hydrogel.

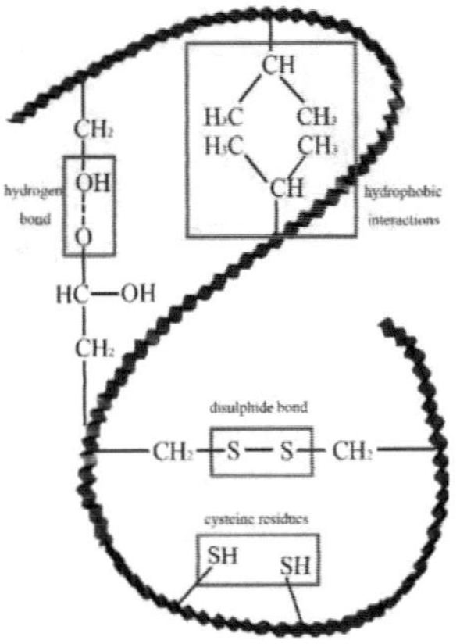

Figure 3. the inter-and intra-molecular bonding in keratin.

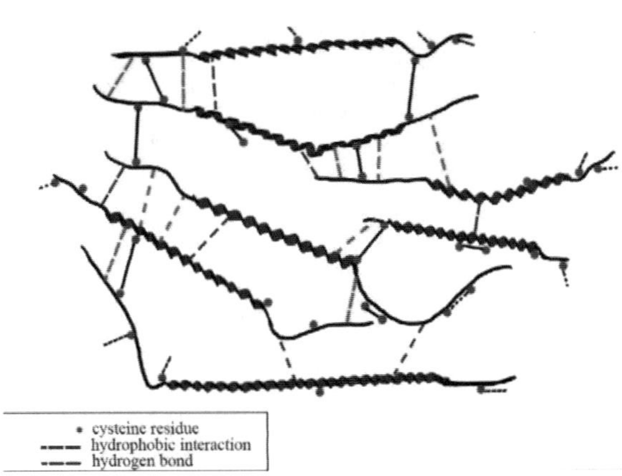

Figure 4. the scheme of proposed self-assembly of rabbit keratin hydrogel.

3.3. Analysis of the swelling ratio

The biggest characteristic of hydrogel is swelling. It shows obvious swelling phenomenon in the solvent, and the molecules in the solvent gradually into the network structure of the hydrogel, the volume expand, but won`t be dissolved. In the Figure 5, the volume of the hydrogel expands because of the molecules of solvent enter the interior of the hydrogel constantly, the chains change and the chains keep moving. When the weight is stable, we call it swelling equilibrium.

Figure 6 shows three stages of the swelling of hydrogels. The first stage is that the molecules of solvent spread to the dried hydrogel, at this point, the hydrogel absorbs few of solvent, filled the internal void constantly, and the volume has not changed much. The second stage is that when the pore structure inside the hydrogel is filled with solvent molecules, the chains inside start to relax and the volume to expand. The third stage is when the molecular chains of the hydrogel start to spread out into the solvent, the structure becomes loose.

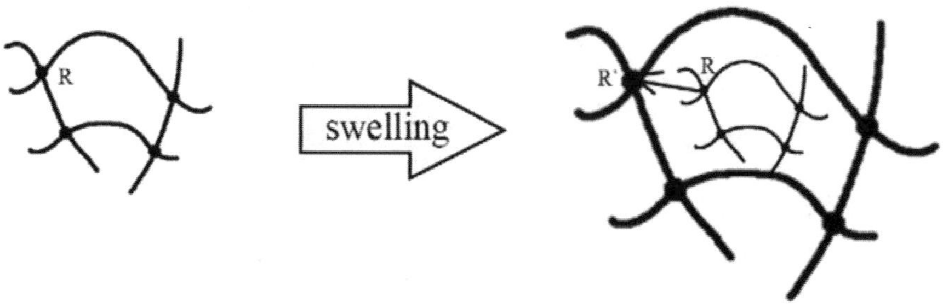

Figure 5. Schematic of keratin hydrogel swelling.

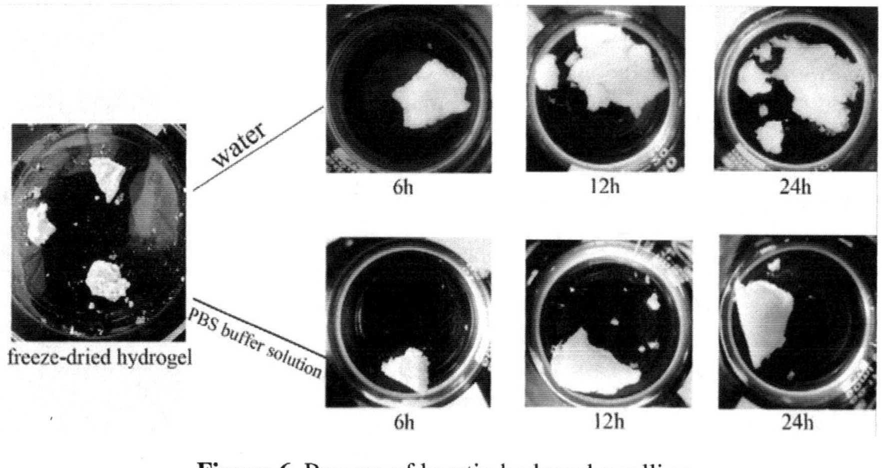

Figure 6. Process of keratin hydrogel swelling.

Figure 7 shows the swelling property of hydrogels with different keratin concentration. When the keratin concentration is about 150mg/ml, the swelling ratio of hydrogel in distilled water is 20.21% and in PBS buffer solution is 9.55%; when is 175mg/ml, the swelling ratio are 14.89% and 7.02% respectively; when is 200mg/ml, there are 10.54% and 3.48%. The structure of hydrogel with low concentration keratin has many large and loose pores, exposes lots of hydrophilic groups, so that water molecules can enter its interior easily and quickly. But with the increase of keratin concentration, the number of cross-link points increasing and the density of cross-link increasing either, the excessive cross-link causes the number of pores decreasing, hinders the penetration of water molecules into the hydrogel, and results the water content ratio and swelling ratio decreasing.

The swelling ratio in distilled water is significantly higher than the swelling ratio in PBS buffer solution. This can be interpreted in electrostatic repulsion and osmotic pressure. On the one hand, a large number of phosphate anions and chloride ions exist in PBS, which have an electrostatic repulsion to the hydroxyl ions inside the hydrogel. The macromolecular chains are stretched to different levels, and then reduce the swelling equilibrium of the hydrogel. On the other hand, osmotic pressure occurs between the inside and outside of the hydrogel because of the differences in the concentration of ions. The osmotic pressure provides power to hydrogel swelling. When the hydrogel swelling in PBS, the cations such as sodium (Na^+) and kalium ion(K^+) entered into the hydrogel network structure, formed a counterbalance with the carboxylic ions, and shielded the sample to avoid external charges. Based on the above-mentioned conditions, the swelling ratio in PBS is lower than it in distilled water.

3.4. TGA analysis

Figure 8 shows TG and DTG traces of the rabbit keratin hydrogel. The weight lost from 50℃ to 200℃ was attributed to the evaporation of absorbed water. The major weight losses took place in the temperature range from 200 ℃ to 450 ℃. It was observed that keratin hydrogel was rapidly decomposed in the temperature range from 200℃ to 450℃ and produced a significant amount of residue. The weight loss specified the degradation of the protein chain molecules, release of Sulphur dioxide and hydrogen sulfide due to breakage of disulfide bond. A peak in the 200–250°C temperature range, related to denaturation peak of α-helix crystallites, and the area under the curve is a measure of the α-helix content. It found that the extracted keratin hydrogel started to decomposed at about 204℃. When the loss residual weight of keratin hydrogel was about 95%, the decomposition temperature was 85℃; when was 50%, the temperature was 342.6℃;when was 25%, the temperature was 597.5℃。

According to the DTG line, it can be seen that the temperature at the maximum point of weight loss slope is 302℃.

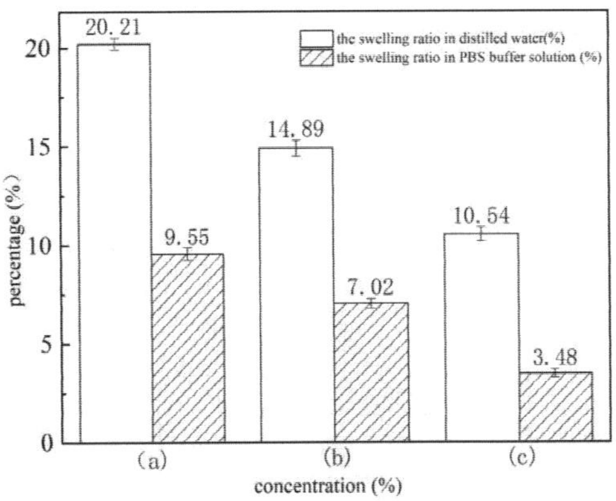

Figure 7. Swelling ratio: (a)low concentration; (b) medium concentration;(c) high concentration.

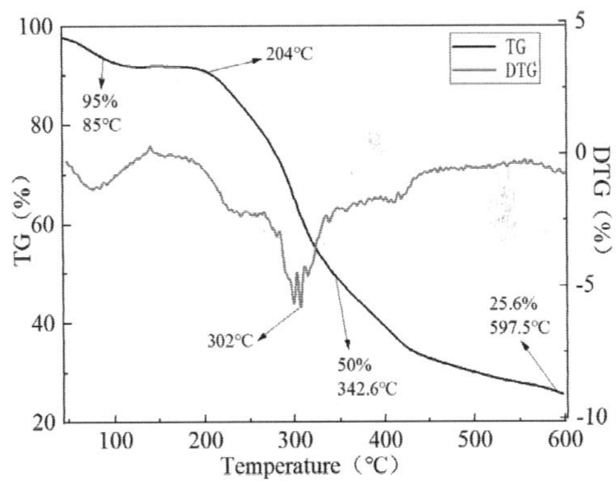

Figure 8. TG and DTG traces.

4. Conclusions

Rabbit hair keratin hydrogel have been prepared by the method of heating-cooling and characterized in terms of SDS-PAGE, SEM, swelling ratio and TG. The effects of keratin addition concentration and temperature on hydrogel formatting ability were studied, under these conditions of experiment, the critical keratin concentration was 125mg/ml, as well as the critical temperature was 55℃, so the conditions of preparing keratin hydrogel are satisfied easily. As revealed by SEM images, the three-dimensional keratin-based hydrogel with a distinctly interconnected porous structure was fabricated successfully. According to the result of swelling assay, the hydrogel material will swell but less dissolve. Such property and special structural characteristic enable the material to retain a large amount of water or biological fluids, so it can be used as raw material in medical dressings and masks. Among the test of TG and DTG, the sample has good thermal stability.

The results demonstrate the correctness and the validity of preparing hydrogel with keratin from waste rabbit hair. Compared with traditional keratin hydrogel materials, rabbit hair has a large production around the world and a lot of short hairs are discarded, businesses can collect and reuse the waste which has low cost and high keratin content. The material is simple in preparative method, controllable in quality and has no need of crosslinking agent. In conclusion, rabbit keratin hydrogel has broad application prospects in the field of medical dressings, body tissue scaffolds, wastewater filters and any other fields, and is worthy of further researching.

Acknowledgments

We would like to thank the present study was supported by China Modern Agriculture Research System (Project NO. CARS-43-E-2) for financial support.

Reference

[1] Sun XX, Bao Y, Peng QY, Chen TY, Lu XY and Yang M 2020 *Materials Reports* **34(07)** 7161-7167

[2] Gao FY, Wei DL, Zhang X, Kou ZL, Wei DP and Lan P 2018 Research progress of hydrogel and its application in biomedicine *New Chemical Materials* **46(S1)** 6-10

[3] Long, YR, Song B, Shi CT, Liu WT and Gu HB 2020 *JOURNAL OF APPLIED POLYMER SCIENCE* **137(19)** 48653

[4] Javeria I, Islam A, Gull N, Ghaffar A, Khan SM, Sabir A, Khaliq S, Butt MTZ, Atta S 2019 Controlled release of Montelukast Sodium from pH-sensitive injectable hydrogels Polymer-Plastics *Technology and Materials* **58(9)** 948-956

[5] Plunkett KN, Berkowski KL, Moore JS 2005 Chymotrypsin re-sponsive hydrogel: application of a disulfide exchange protocol for the preparation of meth acrylamide containing peptides *Biomacromolecules* **6(2)** 632

[6] Xiang YB, Zhang GL, Chen CW, Liu B, Cai DQ, Wu ZY 2017 *ACS Sustainable Chemistry & Engineering* **6(1)** 1192-1201

[7] Kuddushi M, Rajput S, Shah A, Mata J, Aswal VK, El Seoud O, Kumar A, Malek NI 2019 Stimuli Responsive, Self-Sustainable, and Self-Healable Functionalized Hydrogel with Dual Gelation, Load-Bearing, and Dye-Absorbing Properties *ACS Applied Materials and Interfaces* **11(21)** 19572-83

[8] Bai CL, Wang L, Zhu ZY 2020 Adsorption of Cr(III) and Pb(II) by graphene oxide/alginate hydrogel membrane: Characterization, adsorption kinetics, isotherm and thermodynamics studies International *Journal of Biological Macromolecules* **147** 898-910

[9] Karino T, Shibayama M, Ito K 2006 Slide-ring gel: Topological gel with freely movable cross-links *Physical B* **385-386(1)** 692-696

[10] Haraguchi K, Ebato M, Takehisa T 2006 Polymer–Clay Nanocomposites Exhibiting Abnormal Necking Phenomena Accompanied by extremely Large Reversible Elongations and Excellent Transparency *Advanced Materials* **18(17)** 2250-54

[11] Tang J 2019 Developing Tendency and Suggestion of Rabbit Hair Industry *Chinese Journal of Rabbit Farming* **2019(05)** 21-25

[12] Xu YF, Zhang Y 2019 The Invention Relates to a Preparation Method of Graphene Modified Rabbit Hair Fiber *Chinese Journal of Rabbit Farming* **2019(06)** 12-14

Synthesis of cationic graft modified chitosan and its antibacterial finish on rabbit wool fabric

Yu Han Bai , Yi Zhang, Hao Zhang[1] and Xue Lu

School of Textile Science and Engineering, Tiangong University, Tianjin, China

[1]Email:zhanghaoyangyong@163.com

Abstract. As a natural polymer antibacterial agent, chitosan itself has excellent characteristics, but its application is limited due to its poor solubility. Based on this, using ammonium persulfate ($(NH_4)_2S_2O_8$) and sodium bisulfite ($NaHSO_3$) as redox initiators, a cationic antibacterial monomer was grafted onto chitosan to prepare a new type of modified chitosan. The target product was successfully analyzed by infrared spectrum analysis and XPS energy spectrum analysis. Using polycarboxylic acid as a cross-linking agent and modified chitosan as an antibacterial agent, the rabbit wool fabric was finished, and the finished fabric was tested for antibacterial properties and washing resistance. The effect of graft-modified chitosan with different reaction conditions on the antibacterial finishing fabric, such as the ratio of monomer to chitosan substance, reaction temperature, and the amount of initiator added, was studied. The results showed that the finished rabbit wool fabrics had good antibacterial effects on E. coli, and their antibacterial rates reached 100%, and after 20 washings, the antibacterial rate still reached 100%, with long-lasting antibacterial properties. Considering the cost, the ratio of the amount of the chitosan to the monomer was finally selected as 4: 1 to prepare the modified chitosan.

1. Introduction

Textile antibacterial finishing agents are basically divided into inorganic antibacterial agents, organic antibacterial agents, polymer antibacterial agents, etc [1]. As a natural polymer antibacterial agent, chitosan is favored by people because of its excellent characteristics. Compared with general antibacterial agents, it has the advantages of broad spectrum, high antibacterial activity, and low toxicity to mammalian cells [2], but its application is limited to some extent due to its poor solubility [3-4]. The molecular structure of chitosan contains active hydroxyl and amino groups, and functional functional groups can be introduced into the molecular structure of chitosan through chemical modification to obtain chitosan derivatives, which can improve antibacterial properties and other physical and chemical properties and expand the scope of application [5,6,7].

At present, common chitosan modification methods are: quaternization, acylation, carboxylation, graft copolymerization, etc. [8,9,10]. Chitosan's water solubility can be improved after quaternary ammonium salt modification [11]. Quaternization is mainly an amino group at the C2 position of chitosan. The chitosan quaternary ammonium salt can be synthesized by the halogenation method with an alkyl halide, or it can be directly reacted with the quaternary ammonium compound to generate chitosan quaternary ammonium[12]. Acylation modification is the reaction of amino and hydroxyl groups on chitosan macromolecules with different organic acid derivatives such as acid anhydrides and acid chlorides, and the introduction of aromatic or aliphatic acyl groups with different relative molecular weights on their macromolecular chains [13]. Carboxylated chitosan has good water solubility, not only has the advantages of strong antibacterial properties, good freshness, etc. but it is also an amphoteric polyelectrolyte [14-15]. Graft copolymerization is the introduction of alkyl chains, polyether chains, and high polymers on the hydroxyl and amino groups of chitosan [16] to change the molecular structure, the type and number of branches of chitosan to improve its solubility and biocompatibility, and to give it new properties [17,18].

In this study, a new type of reactive cationic antibacterial monomer was synthesized, and the chitosan was modified by graft copolymerization method to form a new type of modified chitosan polymer antibacterial agent. the study. And it was applied to the antibacterial finishing of wool fabrics by cross-linking agents, and the antibacterial properties and wash-ability of the finished fabrics were studied.

2. Experimental

2.1. Materials

Reagents: Dimethylaminoethyl acrylate(DMAEA)and tetradecane bromo were provided by Shanghai civic chemical co. LTD; Tianjin fengchuan chemical reagent technology co., LTD. Provides ammonium persulfate ($(NH_4)_2S_2O_8$), sodium bisulfite ($NaHSO_3$), citric acid, acetic acid and sodium hypophosphite(NaH_2PO_2); Chitosan was provided by Shanghai McLean biochemical technology co., LTD., the deacetylation degree of chitosan was 80.0-95.0%; Beijing aoboxing biotechnology co., LTD. provides peptonechlorine Sodium, beef extract and agar. E. coli (ATCC 29522) was provided by Shanghai luwei technology co., LTD.

2.2. Methods

2.2.1. Synthesis of antibacterial monomer. To a double-necked spherical round-bottom flask equipped with a constant-pressure dropping funnel and a condensing reflux device, added 8.4 mL DMAEA in it. Heated in a constant temperature water bath at 40-45℃, and continuously stirred. Added 13.9mL bromotetradecane, and the reaction was performed for 6 hours. Removed it and cooled it.

2.2.2. Synthesis of Graft Modified Chitosan. 0.15g $(NH_4)_2S_2O_8$ and 0.05g $NaHSO_3$ were dissolved in 10mL of distilled water to make an initiator for later use. To the 100 mL flask added 1g of chitosan, 20 mL of distilled water. Then heated in a constant temperature water bath at 45 ℃, and keep stirring. Added 1 mL of monomer, after 15 mins added the initiator dropwise and finish it within 30 min, and react for 3 h. After the reaction, centrifuged the mixture in ethanol (50%) Three times with High speed centrifuge.

2.2.3. Characterization. FTIR: Nicolet 5700 infrared spectrometer was used for testing. KBr tablets were pressed at a constant temperature of 20℃ and a relative humidity of 65%.

XPS: The elemental composition of the modified chitosan was characterized by X-ray photoelectron spectroscopy (XPS, GENESIS 60S, Gemany).

2.2.4. Finishing process of rabbit fur fabric with modified chitosan. To a flask added 40 mL of distilled water, 2 g of citric acid, 0.5 g of modified chitosan, 0.5 mL of 0.5% acetic acid solution. Heated in a constant temperature water bath at 50℃, and continuously stirred. After all the modified chitosan dissolved, added two pieces of 0.75g rabbit fur fabric (one for finishing and one for washing after finishing), 1g of NaH_2PO_2, 0.25mL of JFC (Primary Alcobol Ethoxylate). For 30mins, rolling them with a residual rate of 50%. Dipping them for 30mins, the second rolling was carried out with a residual rate of 200%. Dried it at 80℃ for 5min, then at 130℃ for 10min, took out it and air-dried.

2.2.5. Performance test of finished rabbit fur fabric. Antibacterial performance: According to GB / T 20944.3-2008 "Evaluation of antibacterial performance of textiles-Part 3 shaking method", the shaking flask method was used to determine the antibacterial performance of E. coli before and after treatment, and the antibacterial rate was calculated according to the following formula:

$$Y=\frac{W_t\text{-}Q_t}{W_t}\times 100\% \quad (1)$$

In the formula: Y-the bacteriostatic rate of the sample; W_t -the concentration of live bacteria in the control flask (CFU/mL); Q_t-the concentration of live bacteria in the flask of the experimental group (CFU/mL).

Washing resistance: For the washing method, refer to "GB/T 20944.3-2008 Textiles for the evaluation of antibacterial properties. Part 3: Oscillation method". Compared with the unwashed finished fabric samples, whether the bacteriostatic rate decreased, and the fastness to washing was determined.

3. Results and discussion

3.1. Reaction principle

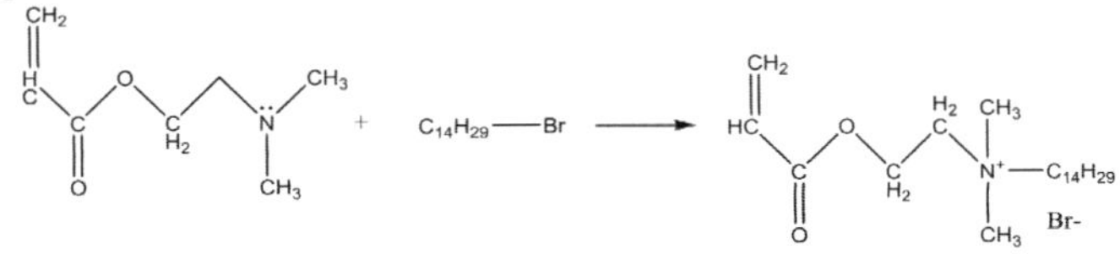

Figure 1. Synthesis process of cationic antibacterial monomer.

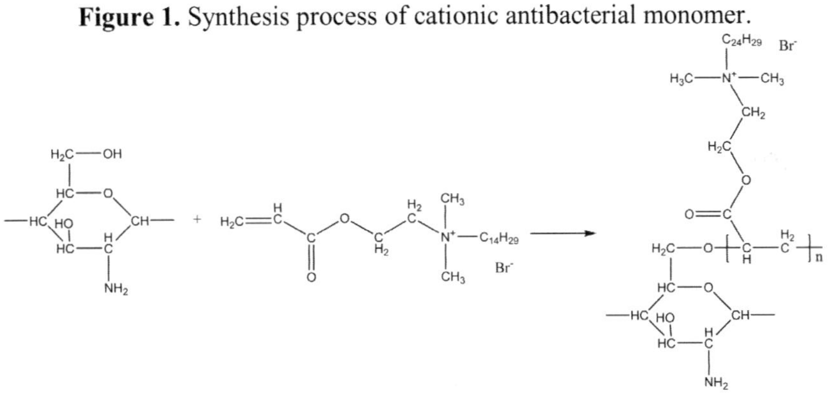

Figure 2. Synthesis process of modified chitosan.

Figure 1 shows the synthesis process of cationic antibacterial monomers. The nucleophilic substitution reaction occurs between dmethylaminoethyl acrylate and bromotetradecane to generate antibacterial monomers with long chain alkyl groups and cationic groups. Figure 2 shows the synthesis process of modified chitosan. The synthesized antibacterial monomer and chitosan undergo graft copolymerization under the initiation of REDOX initiator to synthesize cationic graft chitosan.

3.2. FTIR

Figure 3 is a comparison chart of infrared spectra of modified chitosan and chitosan. It is known that the peak of the carbonyl group is between 1760cm-1-1680cm-1. It can be seen from the figure that the chitosan has no absorption peak between 1760cm-1-1680cm-1, and the modified chitosan has absorption at 1733cm-1. The peak indicates that there is no carbonyl group in the chitosan, but that the modified chitosan contains a carbonyl group, indicating that the carbonyl group is derived from a monomer, indicating that the monomer was successfully grafted onto the chitosan.

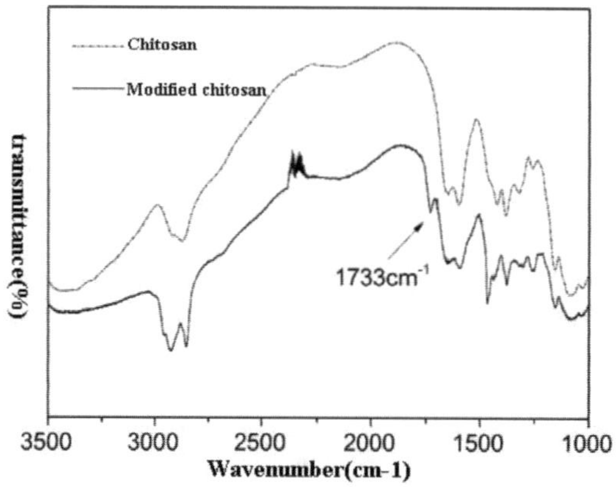

Figure 3. FTIR of modified chitosan and chitosan.

3.3. XPS

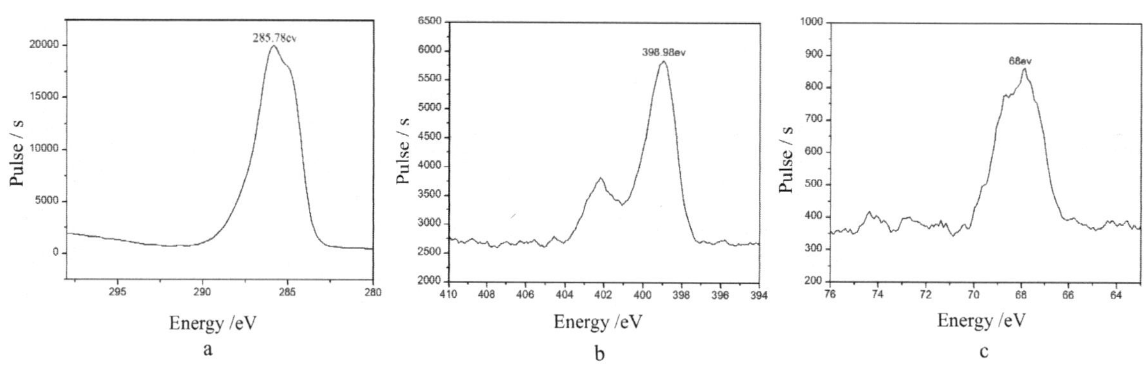

Figure 4. XPS of modified chitosan (a: C atom; b: N atom; c: Br atom).

In Figure 4, 'a' corresponds to the C atomic electron spectrum in the modified chitosan, 'b' to the N atomic electron spectrum in the modified chitosan, and 'c' to the Br atomic electron spectrum in the modified chitosan. As can be seen from figure a, it presents a peak with a binding capacity of 285.78eV, corresponding to the C atom in $MeCH_2NH_2$, which indicates that the C atom in the modified chitosan is connected between the methyl and the N atom, and the N atom is the N atom in the quaternary ammonium salt. In addition, we can also see from other peaks that C is connected to O and H. In figure b, the electron spectral line of N atom in the modified chitosan is corresponding, where the binding energy is 398.98eV peak, and the corresponding N is connected between C atoms. The other peak of binding energy of 401eV corresponds to that the N atom is connected between the ethyl and the halogen atoms, which is the N in the monomer, which also further indicates that the monomer is grafted to the chitosan. In figure c, the XPS electron spectra of Br atom in the modified chitosan showed a peak value of 68eV, indicating that the Br atom was connected to methyl and N atoms. The XPS test results showed that the effective bromine content was 0.48%, and the grafting rate was 2.75% based on the molecular weight of the bromine group.

3.4. Antibacterial performance analysis of rabbit fur fabric

3.4.1. Antibacterial finishing of fabrics with modified chitosan made from different amounts of monomers. Single factor control method was used to change the amount of monomer added. According to the ratios of 2:1, 3:1 and 4:1 between chitosan and monomer, the monomers used were calculated to be 1.3g, 0.85g and 0.65g, and three modified chitosan were prepared, which were denoted as A, B and C. The rabbit wool fabric was treated according to the above finishing technology. Three kinds of chitosan A, B and C were used to conduct antibacterial finishing on the rabbit hair fabric respectively. Other conditions remained unchanged, and the antibacterial test and washing fastness were carried out on the rabbit hair fabric. The results were shown in Table 1 compared with the blank (the antibacterial property of the unfinished fabric).

Table 1. Antibacterial effect of modified chitosan-treated fabrics prepared with different monomer amounts.

Sample	Antibacterial effect(10^{-3})		Bacteriostatic rate Y	
	Not washing	Washing 20 times	Not washing	Washing 20 times
Blank	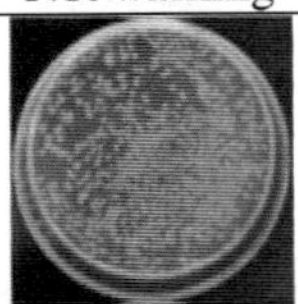	-	-	-
A. 2 : 1	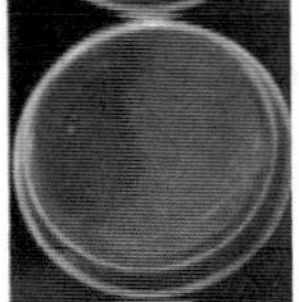	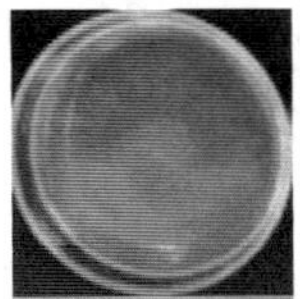	100%	100%
B. 3 : 1	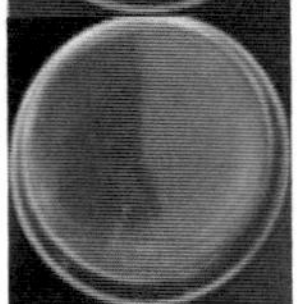	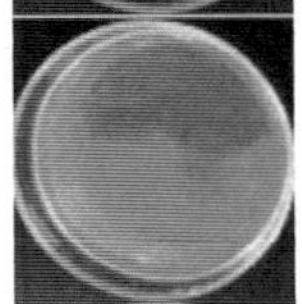	100%	100%
C. 4 : 1	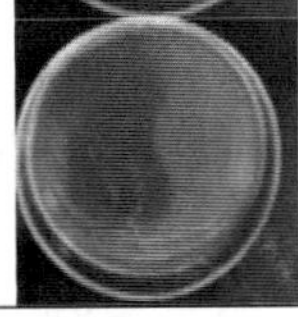	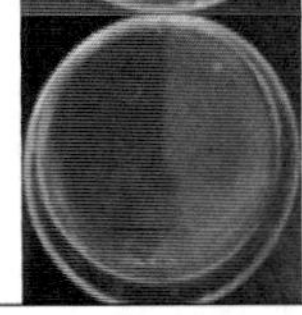	100%	100%

3.4.2. Antibacterial finishing of fabrics with modified chitosan prepared at different reaction temperatures. The single-factor control method was used to change the reaction temperature. The temperatures of the modified chitosan were 40℃, 45℃ and 50℃, and other conditions remained unchanged. Three kinds of chitosan I (40℃), II (45℃) and III (50℃) were used to conduct antibacterial finishing on the rabbit hair fabric respectively. The Antibacterial results were shown in Table 2 compared with the blank (the antibacterial property of the unfinished fabric).

3.4.3. Antibacterial finishing of fabrics with modified chitosan made with different amounts of initiator. Single factor control method was used to change the amount of initiator. Different initiators were added in the preparation of modified chitosan, Other conditions remained unchanged. the chitosan a($(NH_4)_2S_2O_8$:$NaHSO_3$=0.18g:0.06g), b($(NH_4)_2S_2O_8$:$NaHSO_3$=0.3g:0.1g) and c($(NH_4)_2S_2O_8$:$NaHSO_3$=0.45g:0.15g) were used to conduct antibacterial finishing on the rabbit wool fabric, other conditions remained unchanged, and the antibacterial testing and washing fastness were carried out. The results were shown in Table 3 compared with the blank (the antibacterial property of the unfinished fabric).

The single factor control method was used to modify the modification conditions. Three groups of modified chitosan-treated fabrics were tested for antibacterial activity, and the washing group was used as a control. From the number of colonies in the petri dish, the bacteriostatic rate reached 100% in both the washed group and the unwashed group, indicating that the modified chitosan has a good antibacterial effect and has a longer-lasting effect than ammonium. After 20 times washing with standard washing method, the bacteriostatic effect was not affected. Considering the cost, a finishing process with the least amount of monomer is used, that is, the amount of monomer is 0.65g. Modified chitosan can be widely used in various fields and has development value.

Table 2. Antibacterial effect of modified chitosan treated fabrics prepared at different reaction temperature.

sample	Antibacterial effect(10^{-3})		Bacteriostatic rate Y	
	Not washing	Washing 20 times	Not washing	Washing 20 times
blank	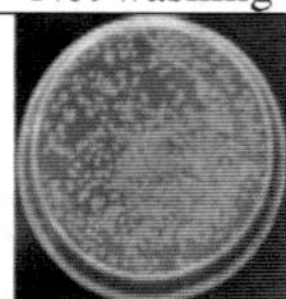	-	-	-
I:40℃	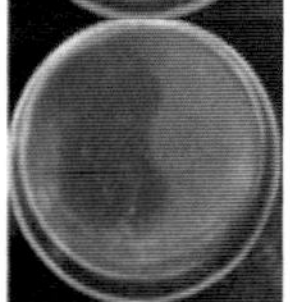	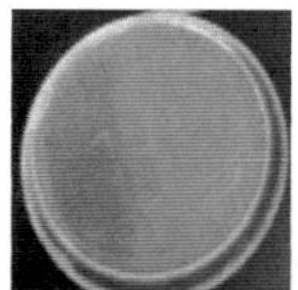	100%	100%
II:45℃	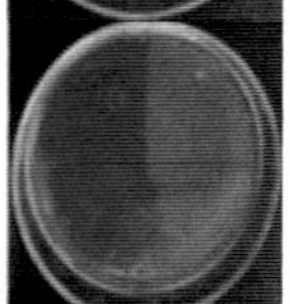	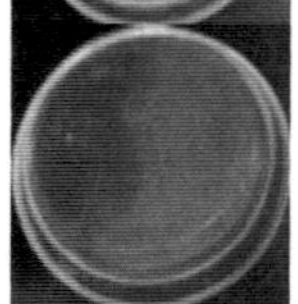	100%	100%
III:50℃	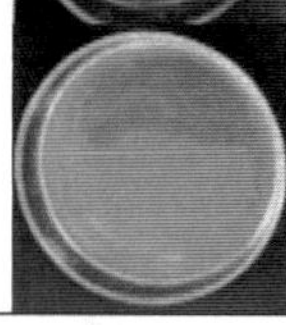	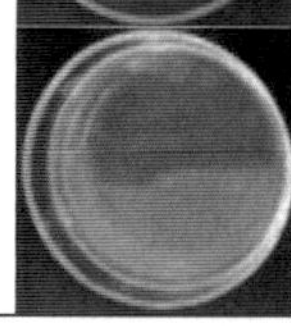	100%	100%

Table 3. Antibacterial effect of modified chitosan treated fabrics prepared with different initiators.

sample	Antibacterial effect(10^{-3})		Bacteriostatic rate Y	
	Not washing	Washing 20 times	Not washing	Washing 20 times
Blank	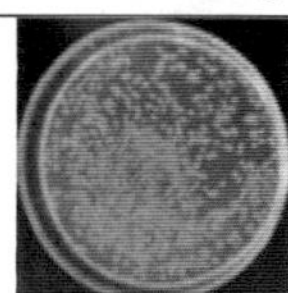	-	-	-
a	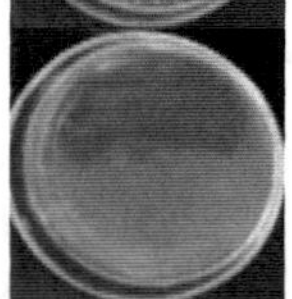	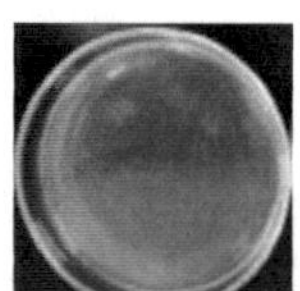	100%	100%
b	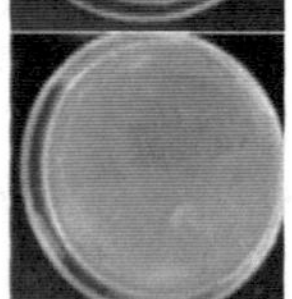	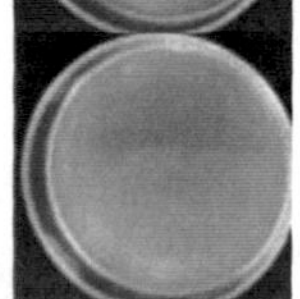	100%	100%
c	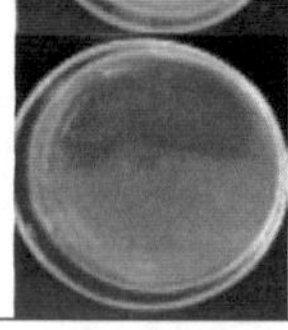	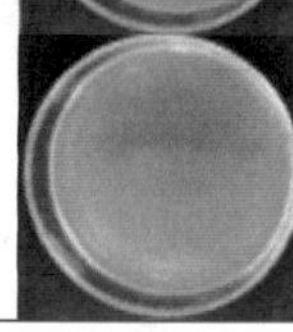	100%	100%

4. Summary

In this study, we used ammonium persulfate ($(NH_4)_2S_2O_8$) and sodium bisulfite ($NaHSO_3$) as redox initiators to graft polymerize cationic antibacterial monomers onto chitosan to prepare a new type of modified chitosan. The grafting rate was 2.75%.

Different modified chitosan was prepared by changing the reaction conditions of graft copolymerization, such as the amount of monomer, reaction temperature, and initiator. Using polycarboxylic acid as a cross-linking agent and modified chitosan prepared under three conditions as bacteriostatic agents, rabbit wool fabrics were finished, and the finished fabrics were tested for antibacterial properties. The results showed that the finished rabbit wool fabric had a good antibacterial effect on E. coli, and the antibacterial rate reached 100%, and after 20 washings, the antibacterial rate still reached 100%, with long-lasting antibacterial properties.

References

[1] Ji Junhui, Shi Weiming 2003 Antibacterial Materials *Chemical Industry Press*
[2] Mo Yuexiang, Luo Jun, Yang Xinhui 2018 Research progress on antibacterial finishing agents for textiles and their detection *Guangdong Chemical Industry* **45 (11)** 150-151
[3] Li Chun, Sun Rong, Zeng Qiuyuan, et al 2011 Preparation and antibacterial mechanism of organic polymer antibacterial agents *Polymer Bulletin* **(3)** 79-85
[4] Ye Yuanli, Li Fei, Feng Zhizhong, Lu Feng, Zhou Li 2018 Research Progress on Antibacterial Finishing of Textiles *Journal of Clothing Industry* **3 (01)** 1-8
[5] Liu Hongli.2017 Synthesis and properties of chitosan derivatives Research *Jiangnan University*
[6] Zhang J, Xia W, Liu P, et al. 2010 Chitosan Modification and Pharmaceutical / Biomedical Applications *Marine Drugs* **8 (7)** 1962-1987
[7] Wang Hao 2017 Research progress and application of modification of chitosan *Journal of Chengdu Textile College* **34 (01)** 187-194
[8] Yu Juan, Xu Junhua, Fan Yimin 2018 Research progress on antibacterial properties of chitosan *Journal of Forestry Engineering* **3(05)** 20-27
[9] GUO Z Y, CHEN R, X1NG R G, et al. 2006 Novel derivatives of chitosan and their antifimgal activities in vitro *Carbohydrate Research* **341** 351-354
[10] Hongkyoon NO, Nayoung PARK, Shinho LEE 2002 Antibacterial Activity of Chitosans and Chitosan Oligoners with Different Molecular Weights *International Journal of Food Microbiology* **74 (1-2)** 65-72
[11] Substituted quaternary ammonium salts of chitosan *Journal of Food Science and Biotechnology*, 2006 **25 (5)** 55-60
[12] Zong Chuanhui, Xu Guoyuan, Sun Yiang, Li Aixiang, Lu Zijian, Li Qiuhong 2017 Recent research progress on the modification of chitosan graft copolymerization *Polymer Bulletin* **05** 19-28
[13] Zhan Wei 2018 Study on Preparation and Application Performance of New Materials Based on Cross-linking Modification of Natural Polysaccharides *Guangxi University*
[14] Chen Jianyong, Liu Guanfeng 2002 Study on Antibacterial Performance of Chitosan Modified and Finished Textiles Using It *Journal of Functional Polymers* **02** 194-198
[15] Zhao Xirong, Li Songlin 2011 Application of tyrosinase in chitosan biological modification. *Chemical Industry and Engineering Progress* **30(11)** 2509-2516
[16] Qin C, Li H, Qi X, etal 2006 Water-solubility of chitosan and its antimicrobial activity. *Carbohydrate Polymers* **63(3)** 367-374
[17] Liu Lunjie, Wu Dayang, Wang Tao 2010 Research Progress of Antibacterial Properties of Chitosan and Development of Antibacterial Textiles *Journal of Textile Research* **31(7)** 145-150
[18] Zhang Zhuanling 2017 Preparation and application of basic amino acid modified chitosan chelating resin *Tianjin Polytechnic University*

Preparation of N, N-dialkyl quaternary ammonium chitosan and its long-lasting antibacterial finishing process for rabbit fur fabric

Yingxue Jiang, Yi Zhang[1] and Hao Zhang

School of Textile Science and Engineering, Tianjin Polytechnic University, Tianjin 300387, China

[1]Email: weiweihanjiangxue@foxmail.com; tianjinzhangyi@126.com

Abstract. N,N-dialkylchitosan (N,N-CTS) was modified by n-butyraldehyde using Schiff base method, it is then synthesized with N,N-CTS by etherification with 3-chloro-2-hydroxypropyltriethylammonium chloride (CHPTAC-ethyl). FTIR and spectrophotometry were used to characterize their chemical structure and physical and chemical properties. The minimum antibacterial activity of the modified chitosan was determined by the minimum bacteriostatic method, and the minimum effective bacteriostatic concentration (MIC) of Escherichia coli was 0.2 g / L, which was better than that of natural chitosan. The antibacterial effect and washing resistance of rabbit wool fabric were investigated by using modified chitosan. After washing for 10 times, the results showed that the inhibition rate of N,N-CCTS on escherichia coli was still more than 99.9%.

1. Introduction

Rabbit fur fabric has good comfort, warmth, moisture absorption and moisturizing effects, and has a natural health effect. However, due to the small amount of impurities in rabbit wool fabrics, there is a large number of microorganisms without special finishing such as washing in the fabric making process, making rabbit wool fabrics more susceptible to microorganisms, and there are few studies on antibacterial finishing of rabbit wool fabrics[1-4].

Chitosan (CTS) is the only biological polymer with amino group in nature. The amino group in the solution dissolved by chitosan is generally easy to be positively charged. By combining with the negative electrons on the cell membrane surface of bacteria, the permeability of the cell membrane of bacteria can be changed so as to inhibit the growth of bacteria. Chitosan is insoluble in water but soluble in weak acids due to intermolecular hydrogen bonding. The amino chemical properties of chitosan are active, and the reaction of chitosan and aldehyde group under acidic conditions produces schiff base, and then the stable n-alkylated chitosan is reduced by sodium borohydride. After the introduction of alkyl group into chitosan molecules, the intermolecular hydrogen bond of chitosan was significantly weakened and the solubility of chitosan was improved. The exposed hydroxyl groups on chitosan also have active chemical properties. The introduction of quaternary ammonium salt group with strong steric hindrance and hydration ability at the hydroxyl group position can also greatly weaken the intermolecular hydrogen bond of chitosan and increase the water solubility of chitosan derivatives[5-8].

Based on this, chitosan was alkylated with n-butyraldehyde, and further modified with etherification agent chptac-ethyl. Finally, N,N-CCTS was obtained. This paper studied the antibacterial activity of N,N-CCTS and applied it to the antibacterial finishing of rabbit fur fabric. The antibacterial activity of rabbit fur fabric before and after finishing was tested, and the feasibility and prospect of N,N-CCTS in the antibacterial finishing of rabbit fur fabric were discussed.

2. Materials and methods

2.1. Materials

Chitosan was purchased from sinopharcine chemical reagent co., ltd. (Shanghai, China), and hydrochloric acid and anhydrous ethanol were provided by Tianjin Fengchuan chemical reagent science and technology co., Ltd.(China). Sodium borohydride and chptac-ethyl were purchased from Shanghai Macklin biochemical co., Ltd. (China), Tianjin Kemiou chemical reagent co., Ltd. (China) provides glacial acetic acid, n-butyraldehyde, sodium lauryl sulfate, sodium hydroxide, sodium hypophosphate, citric acid. The other reagents were of analytical grade.

2.2. Synthesis of chitosan derivatives

The synthesis route of N-CCTS is shown in Figure 1.

Firstly, dissolve chitosan in 5% acetic acid solution, stir at room temperature until dissolved, add n-butyraldehyde of equal molar mass and 2% sodium lauryl sulfate, and stir at room temperature for 8h. Add sodium hydroxide solution to adjust the pH=7. Slowly add 10% $NaBH_4$ solution (1.5 times of aldehyde), stir the reaction for 2h, and then adjust the pH=4.5 with glacial acetic acid. Add the aldehyde a second time and repeat the above reaction step. When the secondary reduction is completed, adjust the pH value to neutral with NaOH solution. Add 70% ethanol to the precipitated product and let it stand for centrifugation. The product is freeze-dried to obtain pure N,N-dialkyl chitosan(N,N-CTS).

Secondly, put a certain amount of alkylated modified chitosan, distilled water, NaOH into the flask, and continue to stir, alkalizate at 45 °C for 30min, and slowly add equal molar mass of CHPTAC-ethyl. After the reaction is completed, the pH is adjusted to neutral with a hydrochloric acid solution. After cooling to room temperature, an equal volume of absolute ethanol is added and the mixture is continuously stirred, and the precipitate is left to stand. After the product is completely precipitated, it is washed with a 50% ethanol solution, centrifuged, and the product is freeze-dried to obtain a relatively pure solid N,N-CCTS.

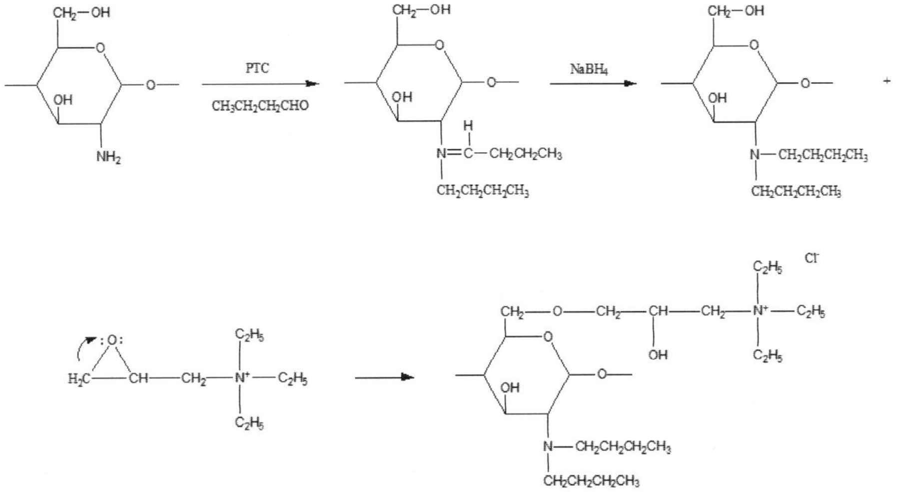

Figure 1. Preparation route of N,N-CCTS.

2.3. Structure and performance characterization
The infrared analysis test of the samples was performed on a Nicolet IS50 infrared spectrometer with a resolution of 4 cm^{-1}, a scanning range of 4000-400 cm^{-1}, and KBr tableting. The sample concentration was 7.5%[9-10].

Determination of the solubility of the sample. Using different concentrations of citric acid solution as a solvent to prepare a 0.5% sample solution, using an ultraviolet spectrophotometer to measure the transmittance of the sample solution at a wavelength of 620 nm[11-12].

2.4. Minimum Inhibitory Concentration
A 0.5% sample solution was prepared by using 0.5% acetic acid solution as a solvent, and the appropriate concentrations of the medium and the appropriate sample solution were mixed to obtain concentrations of 0.05mol/L, 0.1mol/L, 0.15mol/L, 0.2mol/L, and 0.3mol/L, 0.4mol/L, 0.5mol/L mixed medium, after coagulation, 1-2mL of E. coli bacteria suspension was dropped on the plate. Invert the plate in an incubator for 24 hours and observe the growth of the bacteria. The concentration of the test solution that completely inhibits the growth of the colony is the MIC value of the sample to the test bacteria[13-14]. Then prepare the above plate with a mixed concentration of 0.2g / L for sample comparison tests.

2.5. Rabbit fur fabric finishing
Dissolve 2g of finishing agent, 3g of citric acid, and 3g of sodium hypophosphite in 200mL of distilled water, add 3g of rabbit fur fabric, and perform initial soaking at 50 °C for 20min. After soaking, the fabric was taken out for head rolling with the residual rate of 100%, and the fabric was soaked for a second time at 90℃ for 40min. Then the fabric was taken out and rolled for a second time, so that the final liquid rate of the fabric was 200%. After that, it was prebaked at 50℃ for 5min, and then heated up to 90℃ until absolutely dry.

2.6. Determination of antibacterial activity of rabbit fur fabric
GB/T 20944.3-2008 was used to evaluate the antibacterial properties of rabbit fur fabric.

The 0.75±0.05g fabric and the antibacterial treated fabric were respectively loaded into triangular bottles containing the test bacteria solution and exposed to shock at 150rpm and 37℃ for 12h. Take it out, dilute it to an appropriate concentration, spread the plate, and incubate in a biochemical incubator at 37 ℃ for 18 hours. Observe the colony growth of the blank group, control group and test group, respectively. The antibacterial effect was determined based on the results of repeated tests. Bacteriostatic rate is calculated as follows:

$$R = \frac{B-A}{B} \times 100\%$$

Where R is the bacterial inhibition rate, and B and A are the number of viable bacterial cells before and after shaking the bottle[15-16].

3. Results and Discussion

3.1. Structure and performance analysis
FTIR of raw materials and products measured by KBr tableting method are shown in Figure 2. Compared with a, the bending peak of the primary amine N-H of b near 1600 cm^{-1} disappeared, indicating that substitution occurred at $-NH_2$, and the in-plane bending vibration peaks of $-CH_3$ and $-CH_2$ appearing near 1400cm^{-1} and the stretching vibration peaks of C-N of secondary amine at 1160cm^{-1}, indicating that the alkyl group was successfully introduced at the amino position of chitosan to obtain N,N-CTS. Compared with a and b, the C-O stretching vibration peak of c in primary alcohols near 1060cm^{-1} basically disappeared, and the absorption peak of dialkyl ether near 1150cm^{-1} was

enhanced, indicating that the quaternary ammonium group was introduced into chitosan molecules by forming ether bonds with the hydroxyl group on C-6, and N,N-CCTS was finally obtained.

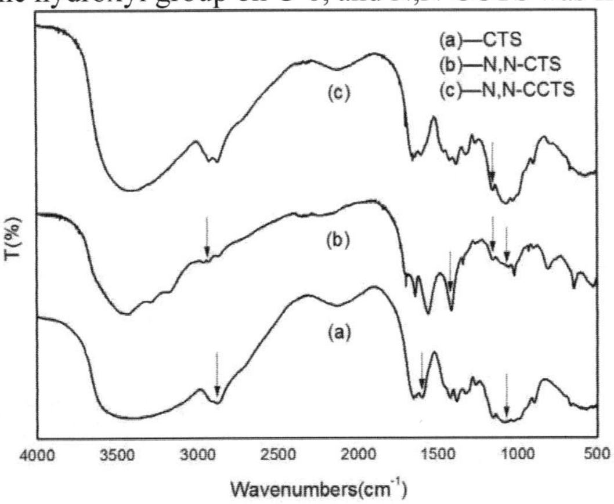

Figure 2. FTIR spectra of raw materials and products.

The test was carried out by analyzing the light transmittance of the chitosan solution. The better the solubility, the greater the light transmittance; on the contrary, the light transmittance was smaller. Figure 3 shows that the transmittance of the samples increases with increasing citric acid concentration. After the introduction of an alkyl group into the chitosan molecule, the intermolecular hydrogen bonding of the chitosan was significantly weakened, and the solubility of the chitosan was improved. Therefore, the solubility of N,N-CTS in weak acids was significantly improved compared with CTS. Then, cationic quaternary ammonium groups were introduced into the alkylated chitosan molecular chain, which destroyed the hydrogen bond between groups of different molecular chains, reduced crystallinity, significantly improved solubility. This shows that the solubility of N,N-CCTS in weak acid is better than that of N,N-CTS and CTS, and it is speculated that the improvement of solubility may be the factor for the improvement of antibacterial performance of modified chitosan.

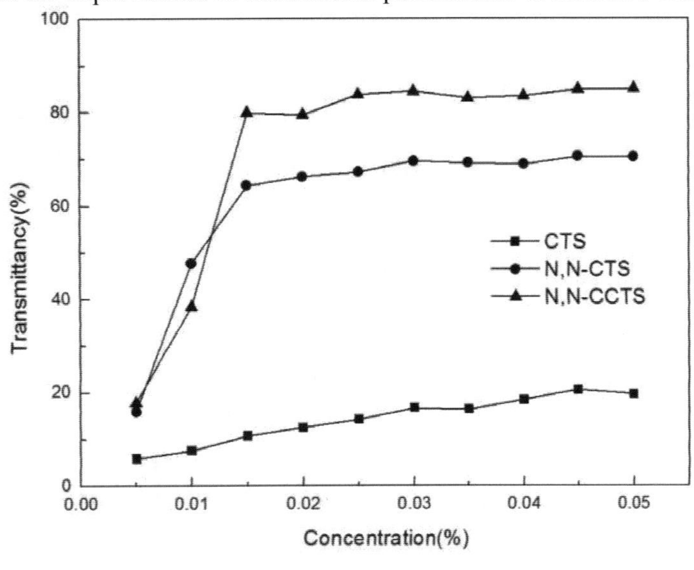

Figure 3. Transmittance of samples at different citric acid concentrations.

3.2. Minimum Inhibitory Concentration

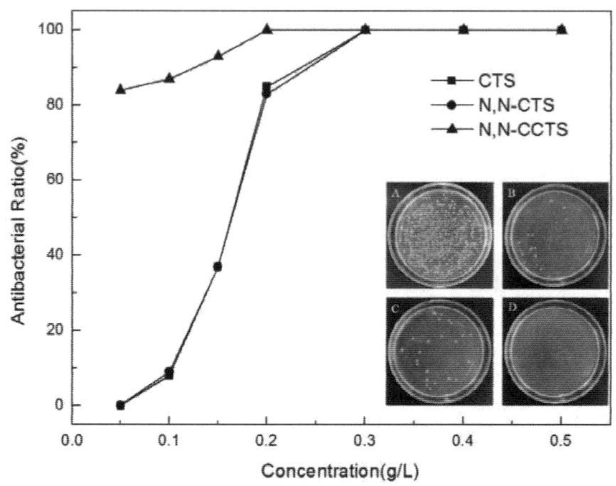

Figure 4. Antibacterial rate and colony growth of the sample against E. coli.
(A - blank group, B - chitosan, C - N,N-CTS and D - N,N-CCTS are the colony growth situation when the sample concentration is 0.2g/L)

Figure 4 clearly shows the results of the sample's promotion, inhibition and bacterial death on E. coli. The positively charged free amino group on chitosan can be combined with the negative charge on the cell wall to form a polymer protective film on the surface of the bacteria to prevent nutrients from being transported into the bacteria. After the initial alkylation modification, the antibacterial properties of the modified chitosan were basically unchanged, and the minimum antibacterial concentration had no significant change compared with chitosan, with MIC values of 0.3g/L. After further modification of N,N-CCTS contains stronger quaternary ammonium groups than amino groups, enhanced bacteriostasis, and the MIC value increased to 0.2g/L. And the solubility is enhanced, and the formed protective film is more dense, preventing the transport of nutrients to the bacteria, and at the same time, causing the cell wall to form holes and causing the leakage of intracellular components, leading to bacterial death. Therefore, the introduction of quaternary ammonium groups further enhanced the antibacterial performance of N,N-CCTS.

3.3. Antibacterial properties of rabbit fur fabric

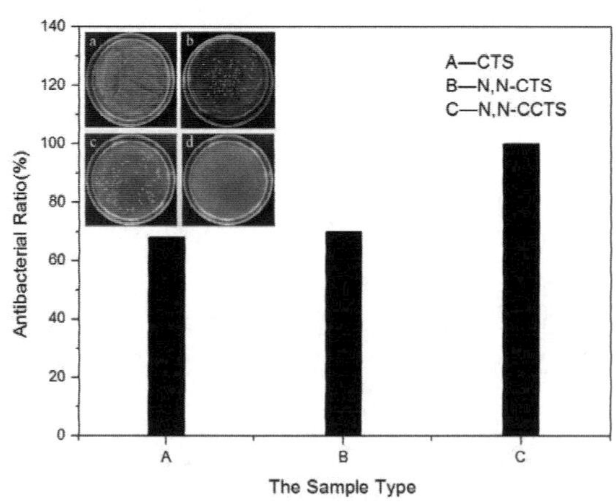

Figure 5. Antibacterial rate of the sample against E. coli.
(a - unfinished fabric, b - CTS finished fabric, c - N,N-CTS finished fabric, d - N,N-CCTS finished fabric were all washed for 5 times for fabric, and the test solution was diluted 10 times)

The comparison of the antibacterial effect of rabbit wool fabrics treated with different finishing agents on E. coli is shown in Figure 5. The results are approximately consistent with the antibacterial rate of the previous section, and the antibacterial rate of modified N,N-CCTS reaches over 99.9%. The amino group and hydroxyl group on the chitosan molecule react with the carboxyl group on the citric acid to form the amide bond and ester bond, which are cross-linked to the rabbit hair fiber. After washing for 5 times, the chitosan has a better bacteriostatic effect. The solubility of N,N-CTS increased, and its bacteriostatic properties did not decrease significantly compared with chitosan.The quaternary ammonium group in N,N-CCTS further improves its solubility and, as a stronger bacteriostatic group, is more prone to cross-linking reactions. Therefore, its washing resistance and bacteriostatic performance are the best.

4. Conclusions

N,N-CCTS has been successfully prepared. The MIC value of N,N-CCTS to E. coli was 0.2g/L, which was better than the MIC value of natural chitosan. The antibacterial activity of rabbit wool fabric treated with N,N-CCTS was significantly improved against E. coli, the bacteriostatic rate is higher than N, N-CTS and natural chitosan, reaching more than 99.9%, which indicates that N,N-CCTS is a long-lasting and excellent antibacterial finishing agent for wool fabrics, and it will have a good application prospect in antibacterial finishing for wool fabrics.

Acknowledgments
The authors gratefully acknowledge the national rabbit industry technology system construction project for funding (Project Number: CARS-44-D-2).

References

[1] Gao Y, Cranston R 2008 Recent advances in antimicrobial treatments of textile *Text. Res. J.* **78** 60-72
[2] Purwar R, Joshi M 2004 Recent developments in antimi-crobial finishing of textiles—a review *AATCC Review* **4** 22-26
[3] Williams J F, HaloSource V and Cho U 2005 Antimicrobial functions for synthetic fibers: recent developments *AATCC Review* **5** 17-21
[4] Mao J W, Murphy L 2001 Durable freshness for textiles *AATCC Review* **1** 28-31
[5] De Arruda I N Q, Pereira V A and Stefani R 2017 Application of chitosan matrix for delivery of rutin *J. Iran. Chem. Soc.* **14** 561-6
[6] Kamel N A, EI-Messieh S L and Saleh N M 2017 Chitosan/banana peel powder nanocomposites for wound dressing application: Preparation and characterization *Mater Sci. Eng. C* **72** 543-50
[7] Ravi K M N V, Muzzarelli R A A, Muzzarelli C, Sashiwa H and Domb A J 2004 Chitosan chemistry and pharmaceutical perspectives *Chem. Rev.* **104** 6071-84
[8] KAWABATA N, NISHIGUCHI M 1998 Antibacterial activity of soluble pyridinium-type polymers *Appl Environ Microbiol* **54(14)** 2532-2535
[9] FanM, Hu Q L and Shen K 2009 Preparation and structure of chitosan soluble in wide pH range *Carbohydr. Polym.* **78** 66-71
[10] Putri R, Dwi N, Putri L S, Putri P S, Tetty N I and Venti S 2016 Preparation and properties of arenga starch-chitosan based edible film *Mater Sci. and Eng.* **107** 012047
[11] CHIVANGKUL T, PENGPRECHA S, PADUNGROS P, Siraleartmukul K, Prasongsuk S and Muangsin N 2014 Enhanced water-solubility and mucoadhesion of n, n, n - trimethyl - n - gluconate - n - homo - cysteine thiolactone chitosan *Carbohydr. Polym.* **108** 224-31
[12] MARTUNS A F, BUENO P V, FOLLMANN H D, Nocchi S R, Nakamura C V, Rubira A F and Muniz E C 2013 Synthesis, characterization, and cytotoxicity of TMC - graft - poly (vinyl alcohol) copolymers *Carbohydr. Res.* **381** 153-60
[13] Marieh G, Mehran G, Sabihe S Z and Sheila B T 2019 Preparing natural biocomposites of N-

quaternary chitosan with antibacterial activity to reduce consumption of antibacterial drugs *J. Hazard. Mater.* **371** 224-232
[14] Angela K T, Ronaldo R F, Lourdes V A, Ursela G B and John P M 2019 Comparative antibacterial activity of commercial chitosan and chitosan extracted from Auricularia sp *Biocatalysis and Agricultural Biotechnology* **17** 189-195
[15] Lidija F Z, Julija V, Tijana R, Matej B, Olivera S and Tatjana K 2013 Antimicrobial and antioxidant functionalization of viscose fabric using chitosan-curcumin formulations *Text. Res. J.* **0(00)** 1-12
[16] Zhang W,Zhou J J and Dai X L 2016 Preparation and characterization of reactive chitosan quaternary ammonium salt and its application in antibacterial finishing of cotton fabric *Text. Res. J.* **0(00)** 1-7

IOP Conf. Series: Materials Science and Engineering 1040 (2021) 012006 doi:10.1088/1757-899X/1040/1/012006

Marine based biomaterial-fish collagen enhances the polarization of human macrophage

Chao Liu and Jiao Sun[1]

Shanghai Biomaterials Research and Testing Center, Shanghai Ninth People's Hospital, Shanghai Jiao Tong University School of Medicine, Shanghai 200023, P.R. China

[1]Email: jiaosun59@126.com

Abstract. Fish collagen have the potential to be used in biomedical fields due to the biocompatible and biodegradable features. However, how hydrolyzed fish collagen regulate human macrophage has not been clearly unraveled. The aim of the current study is to investigate the effects of hydrolyzed fish collagen on the polarization state of human macrophages. The cell viability of human macrophages treated by hydrolyzed fish collagen were determined using CCK-8 assay. The polarization of human macrophages treated by hydrolyzed fish collagen were detected by quantitative real-time polymerase chain reaction (qRT-PCR). it was found that 0.25,0.5 and 1mg/ml hydrolyzed fish collagen did not alter the cell viability of human macrophages. Further, the data showed that hydrolyzed fish collagen inhibited the expression of M1 macrophage marker IL-1β and TNF-α while enhanced the expression of M2 macrophage marker arginase 1(Arg1) and IL-10. Taken together, the results indicated that treatment with hydrolyzed fish collagen result in a significant shift towards an M2 phenotype for human macrophages.

1. Introduction

Collagen has been widely used in cosmetics, pharmaceutical and biomedical fields, the available collagen is primarily obtained from terrestrial animals, such as porcine and bovine[1]. However, mammalian origin collagen present religious constraints and has been linked to a high risk of transmitting certain animal diseases such as foot-and-mouth disease (FMD) and bovine spongiform encephalopathy (BSE) and [2]. Fish collagen has now emerged as a safer alternative to terrestrial collagen due to lack of disease transmission risk [3].

To explore more potential use of fish collagen, recently some groups are focusing on the biological activity of hydrolyzed fish collagen (HFC) [4-6]. It has been found that subcritical water-hydrolyzed fish collagen significantly suppressed LPS-stimulated production of HMGB1(high mobility group box 1) in murine macrophages, along with the reduction of cytosolic translocation of HMGB1[5]. More interestingly, prior studies have indicated that hydrolyzed fish collagen possesses osteoinduction properties [7,8], making it a good candidate as the ingredient of bone implant materials.

Immediately after implantation of a biomaterial, the host acts in a complex cascade of events leading to foreign body reaction mediated mainly by macrophages[9]. Activation of the immune system is an important process for the integration of implantation and wound healing. It is important to elucidate the interaction feature between the biomaterials and the immune related cells. There are two main phenotypes designated proinflammatory M1 macrophage and alternatively activated M2 macrophage [10]. The M1 macrophage is related to the early inflammation phase of wound healing, M1 macrophages produce high levels of pro-inflammatory factors, including interleukin (IL)-6, IL-1β,IL-8 , tumor necrosis factor α (TNF-α) and IL-12, while the M2 macrophage is associated with the late stage of wound healing, M2-type macrophages produce high levels of anti-inflammatory factors such as IL-10 and transforming growth factor-β (TGF-β) in vitro culture conditions [11].

Despite the good biocompatibility of collagen-based biomaterials on fibroblasts[12], these materials are generally associated with the constant and lingering activity of M1 macrophage [13]. Earlier studies have shown that hydrolyzed fish collagen exhibits specific anti-inflammatory effects on mouse macrophages [14,15]. Nevertheless, the effect of hydrolyzed fish collagen on human macrophage are little known. The current study was aimed to evaluate the effect of hydrolyzed fish collagen on the cell behavior of human macrophages. Cell viability and polarization state were investigated in human macrophages in this study.

2. Materials and methods

2.1. Materials
Hydrolyzed fish collagen (HFC) was generously provided by the Shanghai Fisheries Research Institute. HFC has a molecular weight distribution of 700 to 1300 Da, the contact angle of HFC is around 26°, There are many kinds of amino acid, such as glycine (333/1000 residues), proline (115/1000 residues), and hydroxyproline(117/1000 residues) [8].

2.2. Cell culture
The human THP-1 macrophage cell line was purchased from Cell Bank of Chinese Academy of Sciences (Shanghai, China). Human macrophage was cultured in RPMI 1640 medium (Gibco, Life Technologies, Carlsbad, CA) containing 10% fetal bovine serum (FBS, Hyclone, USA), 100 IU/ml penicillin (Gibco, USA), and 100 μg/ml streptomycin (Gibco, USA). The cells were maintained at 37 °C in a 5% CO_2 atmosphere, and the medium was replaced every 3 days. To induce the differentiation into macrophage, the human THP-1 macrophages were added into 96-well plates at a density of 1.0×10^5/mL, then the cells were treated with 100 ng/ml PMA (phorbol 12-myristate 13-acetate) (EMD Biosciences, USA) for 48 hours at 37℃. To generate the polarized phenotypes, cells were then cultured for 24 hours in M1 polarization inducing media (100 ng/ml LPS) to generate M1 macrophages, or M2 polarization inducing media (20 ng/ml of IL-4, and 20 ng/ml of IL-13) to generate M2 macrophage.

2.3. CCK-8 assay
The cell viability of human macrophages was evaluated using CCK-8 assay (Dojindo, Kumamoto, Japan) according to the manufacturer's protocol. Briefly, cells were seeded onto 96-well plates with a density of 5×10^4 cells/well, and treated with various concentration of HFC (0, 0.25, 0.5, 1, 2 and 4 mg/ml) for 24 h. Then, the medium was removed by aspiration and 100 μl of medium containing 10% CCK8 reagent was added to each well, after incubation at 37 °C incubator for 2 h, the optical density was read at 450 nm. 3 biological replicates were used in CCK-8 assay.

2.4. Real-time PCR analysis
Based on the results of CCK-8 assays, 0.25 mg/ml HFC was used in this part. M0 human macrophage were cultured for 1 day with 0.25 mg/ml HFC to investigate macrophage polarization. Total RNA extraction and purification were performed using TRIzol reagent (ThermoFisher, USA) according to manufacturer's instruction. cDNA was synthesized from 500ng of total RNA using the Prime Script RT reagent kit (TaKaRa, China) according to the manufacturer's protocol. Real-Time PCR was carried out by using SYBR green (PowerUp SYBR Green Master Mix, Life Technologies, USA) according to manufacturer's protocol. The PCR mixture was performed in a 25 μl that consists of the following: 2 μl of cDNA template, 12.5 μl of 2× SYBR Green, 1 μl of forward primer, 1 μl of reverse primer, and

8.5 μl of double distilled water(ddH2O). The ΔΔCt method was used to calculate the gene expression levels after normalization according to the expression of GAPDH. 3 biological replicates were used in Real-time PCR analysis. The primer sequences were shown in Table 1.

Table 1. Primers for real-time PCR.

Target gene	Forward primer sequence (5'-3')	Reverse primer sequence (5'-3')
IL-1β	ATGATGGCTTATTACAGTGGCAA	GTCGGAGATTCGTAGCTGGA
TNF-α	ACTTTGGAGTGATCGGCC	GCTTGAGGGTTTGCTACAAC
Arg 1	ACGGAAGAATCAGCCTGGTG	GTCCACGTCTCTCAAGCCAA
IL-10	GCTGGAGGACTTTAAGGGTTAC	GTAGTCTGGGTCTTGGTTCTC
GAPDH	AGCCACATCGCTCAGACA	GCCCAATACGACCAAATCC

2.5. Statistical Analysis

All experiments were performed at least three times, Date were expressed as mean ±SEM. Data analysis was carried out using one-way analysis of variance and Student's *t*-test. p values less than 0.05 were considered as statistically significant in all experiments.

3. Results

3.1. The effects of different doses of HFC on cell viability of human THP-1 macrophage

CCK-8 assays were used to determine the optimal concentration of HFC on human macrophage. Human macrophage was exposed to various concentrations of HFC for 24 h. Figure 1 demonstrates that HFC at concentrations 0.25 mg/ml, 0.5 mg/ml and 1 mg/ml did not significantly change cell viability in contrast to cells without addition of HFC, suggesting that 0.25, 0.5 and 1 mg/ml HFC were relatively safe for human macrophage (Figure 1). 0.25 mg/ml is the minimum concentration that did not significantly change cell viability in our experiment, based on the principle of minimum dose, 0.25 mg/ml HFC were used for subsequent experiments.

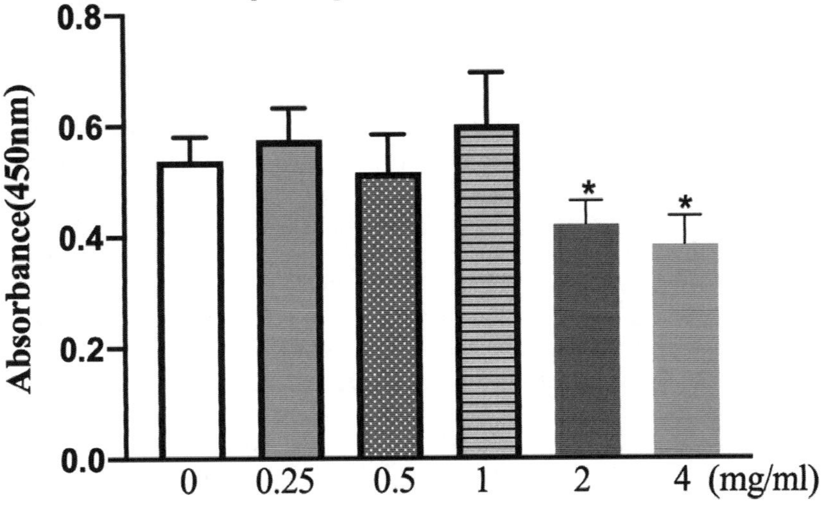

Figure 1. Cell viability of human macrophage treated with different concentrations of HFC. * indicates a significant difference compared with control, $p < 0.05$.

3.2. Macrophage polarization in response to the HFC

The expression of macrophage polarization markers was evaluated by real-time PCR when M0 human macrophage were cultured with 0.25 mg/ml HFC. Both IL-1β and TNF-α, the two important M1 macrophage markers were significantly decreased in HFC group as compared to that in control group (Figure 2A and B), while higher mRNA levels of M2 markers, including Arg1 and IL-10, were observed in cells treated with HFC when compared with control group (Figure 2C and D). It did show evidence that HFC induced macrophage to become polarized toward the M2 phenotype.

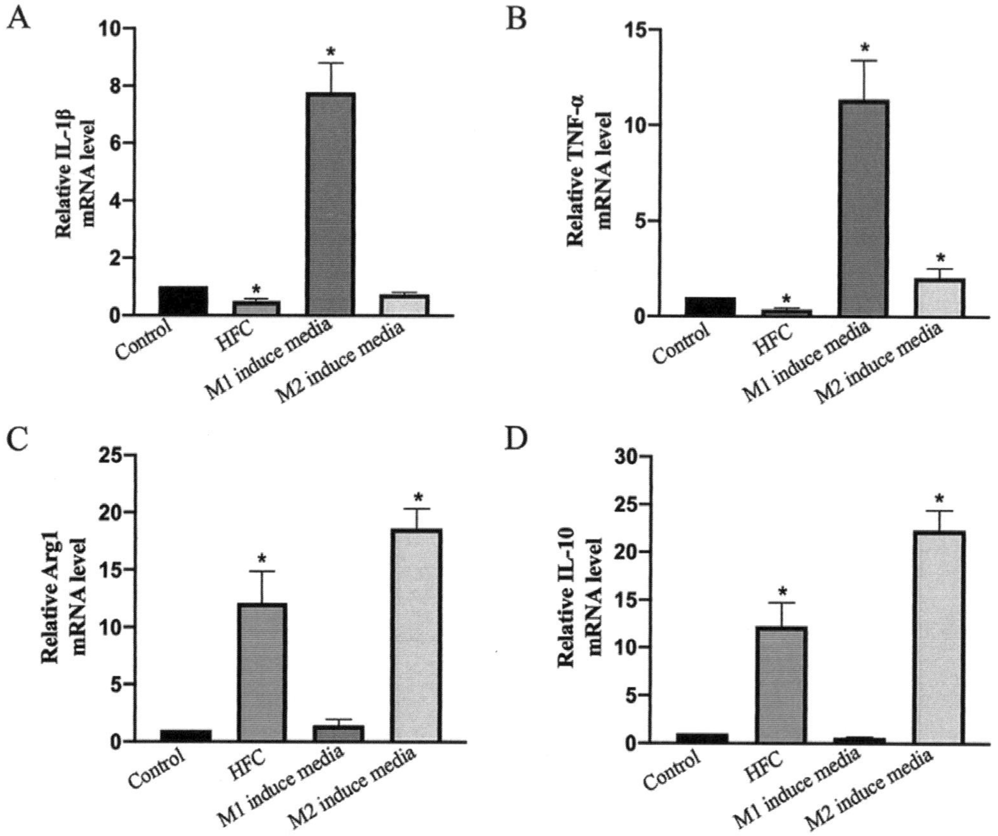

Figure 2. qRT-PCR analyses of IL-1β(A) TNF-α(B), Arg1(C) and IL-10(D) gene expression. *indicates a significant difference between HFC treated groups and control group with p-value < 0.05.

4. Discussion

Amino acids (Proline and Hydroxyproline) are key players in upholding the function of the hydrolyzed collagen. Proline and hydroxyproline are also the most abundant amino acid in mammalian collagen[16], in the present study, since the obtained amino acid composition of HFC was matching to the mammalian, HFC can be definitely used as an alternative to existing commercial collagen, furthermore, it has been shown that HFC possess good bioactivity[17,18].Therefore, HFC is expected to be a good candidate for the biomedical application.

In the present study, 0.25, 0.5 and 1 mg/ml HFC did not alter the cell viability of human macrophage, which suggest the good biocompatibility of HFC at specific concentrations. On the other hand, the recruitment and proliferation of macrophages on the foreign body are triggered by chemokines and cytokines released by the initially adhered macrophages to the implant, which indicates that HFC could potentially serve as a good implant coating material, no interfering the cell viability of macrophages is one of the main characteristics for a good biomaterial[19], but further studies are needed to corroborate this conclusion.

Furthermore, the polarization of human macrophage treated with HFC was evaluated by qRT-PCR. Human macrophage treated with 0.25 mg/ml HFC showed lower expression level of M1 markers (IL-1β and TNF-α), whereas higher expression level of M2 markers (Arg1 and IL-10) compared with that of control. It is commonly believed that the M1 macrophage producing proinflammatory cytokines and associated with tissue disruption, while the M2 macrophage is anti-inflammatory and linked to tissue repair [20]. M1 macrophages with increased TNF-α, IL-1 expression suggests delayed tissue healing and more severe inflammatory response [21,22]. IL-1β caused an activation of both canonical and noncanonical NF-κB pathway, leading to an increase in intestinal epithelial tight junction permeability [23]. Other group found that IL-1β can cause the destruction of junctions of endothelial cells [24]. Tumor necrosis factor-α (TNF-α) is a cytokine mainly produced by macrophages and monocytes, and plays a central role in inflammatory osteoclastogenesis [25]. Therefore, repression of inflammatory cytokine produced by HFC in our study would be helpful to endothelialization and tissue repair.

In addition, HFC induced M2 macrophage polarization by increasing the level of Arg1 and IL-10. Both Arg1 and IL-10 are anti-inflammatory factors with ability to improve tissue regeneration. Arginase 1 is an enzyme of the urea cycle and join in the regulation of a multitude of cellular processes like cell growth, DNA replication and protein translation [26]. IL-10 can suppress the production of the pro-inflammatory cytokines and inhibit the secretion of nitric oxide [27]. Both Arginase 1 and IL-10 are helpful to implant integration and Integrity. Thus, our results demonstrated that HFC inhibited pro-inflammatory cytokine production by macrophages while promoted the production of anti-inflammatory cytokines, HFC contribute to the development of an anti-inflammatory microenvironment that is favorable to tissue repair and regeneration.

5. Conclusions

In this study, we found that a novel marine biomaterial hydrolyzed fish collagen induced the shift of human macrophage phenotype from M1 to M2, which may lead to the application of hydrolyzed fish in biomedical fields.

Acknowledgements

This study was supported by National Natural Science Foundation of China [NO.31600760]. We thank prof. Nanping Wang for providing HFC.

References

[1] Ng WL, Chua CK, and Shen YF 2019 *Progress in Polymer Science* **97** 101145
[2] Swatschek D, Schatton W, Kellermann J, Müller WE and Kreuter J 2002 *Eur. Eur. J. Pharm. Biopharm.* **53** 107-13
[3] Safandowska M and Pietrucha K 2013 *Int. J. Biol. Macromol.* **53** 32-7
[4] Güz BC, Molenaar R, de Jong IC, Kemp B, van den Brand H and van Krimpen M 2019 *Poult. Sci.* **98** 6552-6563
[5] Ahn MY, Hwang JS, Ham SA, Hur J, Jo Y, Lee S, Choi MJ, Han SG and Seo HG 2017 *Biomed. Pharmacother* **93** 923-30
[6] Lee EJ, Hur J, Ham SA, Jo Y, Lee S, Choi MJ and Seo HG 2017 *Int. J. Biol. Macromol.* **104(Pt A)** 281-86
[7] Liu C and Sun J 2015 *Biomed. Mater.* **10** 065020
[8] Liu C and Sun J 2014 *Biomacromolecules* **15** 436-43
[9] Anderson JM, Rodriguez A and Chang DT 2008 *Semin. Immunol.* **20** 86-100
[10] Miron RJ, Chandad F, Buser D, Sculean A, Cochran DL and Zhang Y 2016 *J. Periodontol.* **87** 91-9
[11] Yunna C, Mengru H, Lei W and Weidong C 2020 *Eur. J. Pharmacol.* **29** 173090
[12] Ng WL, Goh MH, Yeong WY and Naing MW.2018 *Biomaterials science* **6** 562-574
[13] Brown BN, Londono R, Tottey S, Zhang L, Kukla KA, Wolf MT, Daly KA, Reing JE and Badylak SF 2012 *Acta Biomater.* **8** 978-87
[14] Liu C, Liu X, Xue Y, Ding T and Sun J. 2015 *RSC advances* **5** 30727-30736
[15] Liu C and Sun J. 2019, *Life Sci.* **238** 116970
[16] Capella-Monsonís H, Coentro JQ, Graceffa V, Wu Z and Zeugolis DI. 2018 *Nat Protoc.* **13** 507-529

[17] Liu C, Xue Y and Sun J 2014 *Advanced Materials Research.* **1025-1026** 570-573
[18] Liu C and Sun J. 2020 Experimental and Therapeutic Medicine. **20** 375-384
[19] Malanagahalli S, Murera D, Martín C, et al. 2020 *Nanomaterials (Basel).* **10** 228
[20] Funes SC, Rios M, Escobar-Vera J and Kalergis AM. 2018 *Immunology* **154** 186-195
[21] Tarique AA, Logan J, Thomas E, Holt PG, Sly PD, and Fantino E. 2015 *Am. J. Respir. Cell. Mol. Biol.* **53** 676–88
[22] Jung SH, Hwang JH, Kim SE, Kim YK, Park HC and Lee HT. 2017 *Xenotransplantation* **24**
[23] Al-Sadi R, Ye D, Said HM and Ma TY.2010 *Am. J. Pathol.* **177** 2310-22
[24] Tossetta G et al. 2014 *Placenta* **35** 509–16
[25] Kitaura H, Kimura K, Ishida M, Kohara H, Yoshimatsu M and Takano-Yamamoto T. 2013 *Clin. Dev. Immunol.* **2013** 181849
[26] Rath M, Müller I, Kropf P, Closs EI and Munder M. 2014 *Front. Immunol.* **5** 532
[27] Ameredes BT, Zamora R, Gibson KF, Billiar TR, Dixon-McCarthy B, Watkins S and Calhoun WJ. 2001 *J. Leukoc. Biol.* **70** 730-36

Corrosion behavior of C-22, Ti-0.2Pd and Ti-32Mo alloys in the aqueous solution of hydrochloric acid and sodium chlorate

Mingyue Zhang[1], Bingqing Ma, Jian Xu, Huangwei Zhou, Yi Liu, Zhibing Chen , Ziwei Xu, Jia Chen and Jiafan Ye

Chengdu Industry and Trade College, Chengdu Technician College, Chengdu, Sichuan, P.R. China

[1]Email: zhangmyanhb@126.com

Abstract. Corrosion behavior of Hastelloy C-22, Ti-0.2Pd and Ti-32Mo alloys in the aqueous solution of hydrochloric acid and sodium chlorate was investigated. The C-22 alloy was pitted because the intergranular carbides rich in Cr and Mo would lead to the lack of corrosion-resistant elements in the adjacent area. The Ti-0.2Pd alloy was pitted because the passivation trend decreased in reducing solution and Cl^- broke through the passivation film. The weigh loss rate and corrosion depth of Ti-0.2Pd alloy increased sharply after pitting corrosion, which was due to the rapid corrosion of the materials exposed inside by the corrosion medium. The Ti-32Mo alloy was pitted due to the transpassivation of Mo caused by $NaClO_3$. In the process of corrosion, the products of reaction between Mo and Cl^- formed an insoluble and dense protective film, which slowed down the corrosion rate of Ti-32Mo, that is why the weigh loss rate and corrosion depth increased slowly even if pitting corrosion happened.

1. Introduction

Hastelloy alloy C-22, titanium alloys Ti-0.2Pd and Ti-32Mo are corrosion resistant materials which may be applied in mechanical engineering and chemical apparatus construction [1] [2] [3]. In C-22 alloy, the alloying element Cr is used for corrosion resistance in oxidizing environment and Mo for reducing environment [4]. Therefore, C-22 alloy has good corrosion resistance in both oxidizing and reducing solutions. However, in order to ensure the good corrosion resistance in the environment containing Cl^-, the content of Cr, Mo and W in the C-22 alloy should reach the required value [5]. For Ti-0.2Pd alloy, the addition of Pd increases the exchange current density of Ti and therefore, facilitates passivation in many systems [6]. However, the pitting sensitivity of Ti-0.2Pd increases with the increase of chloride concentration [7]. Ti-32Mo alloy has excellent corrosion resistance in strong reducing medium. However, the anode potential will increase with the addition of strong oxidant, which will lead to transpassivation and corrosion of the alloy[8].

A new type of disinfectant for tap water containing high concentration hydrochloric acid and sodium chlorate has been widely used in recent years. A company once used C-22 alloy as the storage tank material of this disinfectant, while the tank was severely corroded after a period of time. Therefore, it is necessary to find a more suitable corrosion-resistant material in order to store the disinfectant safely.

In this work, the corrosion behavior of C-22, Ti-0.2Pd and Ti-32Mo in hydrochloric acid and sodium chlorate solution was investigated. After a period of corrosion, the weigh loss rate and corrosion depth were measured. Moreover, the macro and micro morphologies of the specimens after corrosion were observed, and the corrosion mechanism was analyzed. This work can provide a reference for the selection of materials used in the environment containing both oxidizing and reducing media.

2. Experimental procedure

Commercially Hastelloy alloy C-22, Ti-0.2Pd and Ti-32Mo alloys were prepared with 25 mm in diameter and 10 mm in length. Compositions of these alloys are listed in Table 1.

The surfaces of all the specimens were ground using 2000# emery paper and ultrasonically cleaned with ethanol for 15 min. Then the specimens were immersed in corrosive solution, which is composed of $NaClO_3$ solution（the mass ratio of $NaClO_3$ to H_2O is 1:2.5） and HCl with the volume ratio of 1:2. After immersed for 2,4,6,8 months, the specimens were taken out for corrosion resistance study.

Table 1. Chemical compositions of C-22, Ti-0.2Pd and Ti-32Mo alloys (wt.%).

Alloy	C-22	Ti-0.2Pd	Ti-32Mo
Pd	--	0.2	--
Mo	13	--	32
Fe	3	0.05*	0.3*
O	--	0.15*	0.015*
N	--	0.007*	0.015*
H	--	0.01*	0.015*
C	0.015*	0.08*	0.1*
Ti	--	Bal.	Bal.
Cr	22	--	--
W	3	--	--
Co	2.5*	--	--
V	0.35*	--	--
S	0.02*	--	--
Si	0.08*	--	0.15*
Ni	Bal.	--	--

* Maximum

In order to study the weigh loss rate, the specimens before and after corrosion were weighed by a precision electronic balance (FA 3204). The macroscopic morphology of the specimens was photographed by a camera (D3300, Nikon). The surface micromorphology and corrosion depth of the specimens were characterized by a color three-dimensional (3D) laser scanning microscope (VK-9710K).

3. Results and discussion

3.1. Weigh loss rate

Figure 1 shows the weigh loss rate of different alloys as a function of corrosion time. The weigh loss rate of all specimens increases with the increase of corrosion time, but the speed of increase was different. After corrosion for two months, the weigh loss rate of Ti-0.2Pd is the lowest, and that of C-22 and Ti-32Mo alloy is close to each other. After corrosion for 4 or 6 months, the weight loss rate was Ti-0.2Pd, Ti-32Mo and C-22 from low to high. After corrosion for 8 months, the weigh loss rate of Ti-0.2Pd increases sharply and becomes much higher than the other two alloys. These phenomena will be analyzed in detail later.

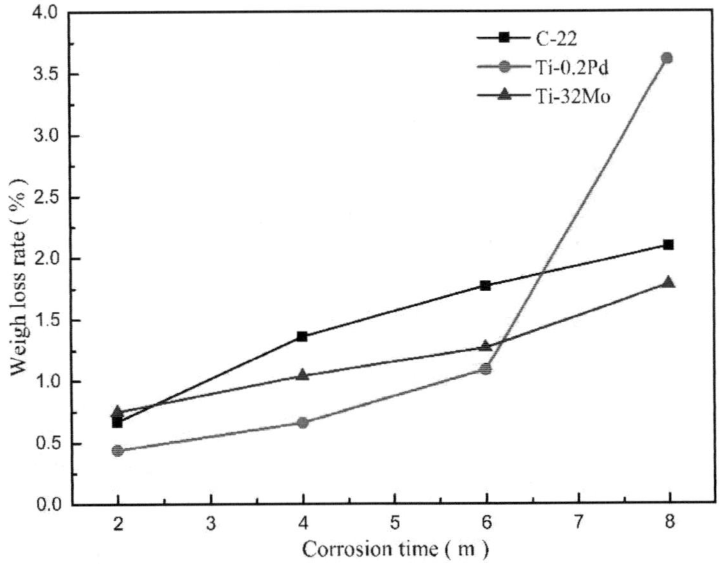

Figure 1. Weigh loss rate of specimens after different corrosion time.

3.2. Macroscopic morphology and corrosion depth

The macro-morphologies and 3D micro-morphologies of specimens after corrosion are shown in Figure 2-4. The corrosion depths are shown in Figure 5. For all of the alloys, the corrosion depth increases with the increase of corrosion time and pitting corrosion occurs after a certain time.

For C-22 alloy, a rough surface and patina is presented after corrosion for 2 months. Pits with a maximum depth of 298.4 μm are observed after corrosion for 4 months. Then locally connected corrosion pits with a maximum depth of 366.4 μm occur after corrosion for 6 months. Finally, a corrosion pit with a diameter of about 5000 μm and a depth of 3000 μm appears and small corrosion pits are scattered around.

For Ti-0.2Pd alloy, it shows uniform corrosion when the corrosion time is less than 4 months, and the corrosion depth after 2 and 4 months is 163.2 μm and 208.3 μm respectively. Scattered pits with a maximum depth of 254.4 μm appear after corrosion for 6 months. Then several large and deep pits with the maximum diameter of about 3000 μm and the depth of 2500 μm occur and small corrosion pits are scattered after corrosion for 8 months.

For Ti-32Mo alloy, it shows uniform corrosion when the corrosion time is less than 4 months, and the corrosion depth after 2 and 4 months is 208.3 μm and 230.3 μm respectively. There are scattered or even flaky corrosion pits with a maximum depth of 400.4 μm after corrosion for 6 months. Then more small pits can be observed and local spalling occurs, and the maximum corrosion depth is 558.8 μm after corrosion for 8 months.

It can be seen from Figure 5 that after the pitting corrosion, the pit depth of C-22 alloy increases slowly at first, and then increases sharply after corrosion for 6 months; the pit depth of Ti-0.2Pd increases sharply, while that of Ti-32Mo increases slowly in the experimental time.

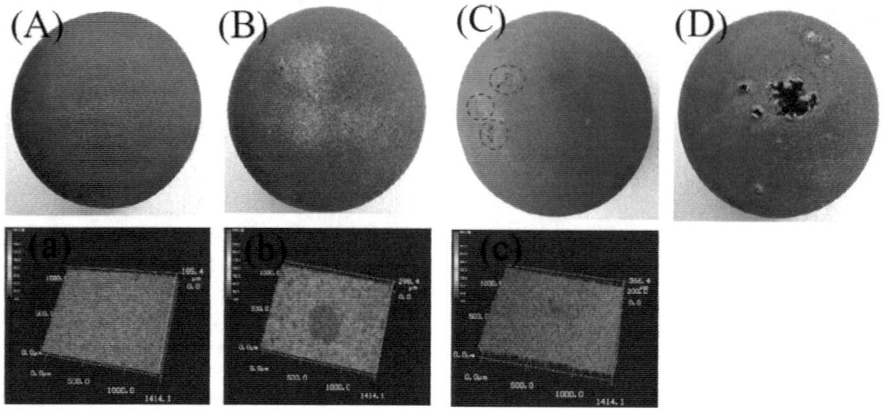

Figure 2. Macro-morphologies and 3D micro-morphologies of C-22 alloy after corrosion for different time: (A, a) -2 months, (B, b) - 4 months, (C, c) - 6 months and d-8 months.

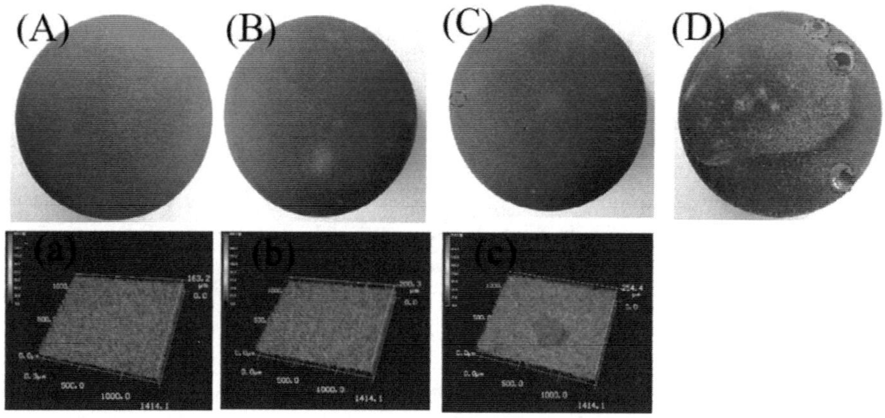

Figure 3. Macro-morphologies and 3D micro-morphologies of Ti-0.2Pd alloy after corrosion for different time: (A, a) -2 months, (B, b) - 4 months, (C, c) - 6 months and d-8 months.

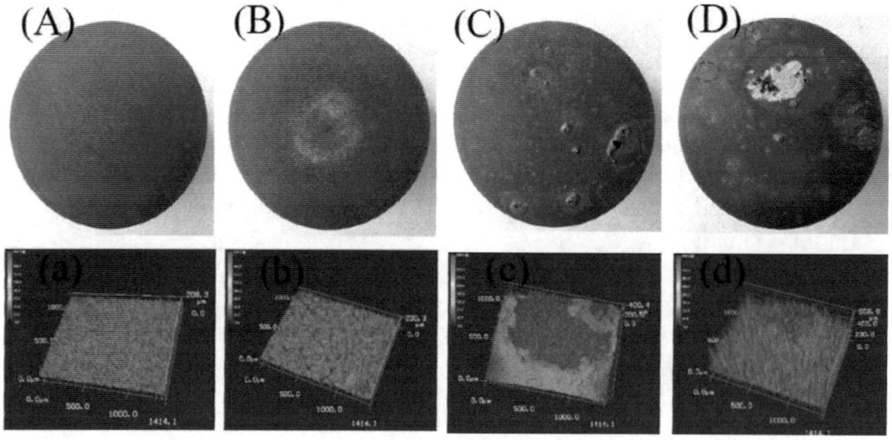

Figure 4. Macro-morphologies and 3D micro-morphologies of Ti-32Mo alloy after corrosion for different time: (A, a) -2 months, (B, b) - 4 months, (C, c) - 6 months and (D, d)-8 months.

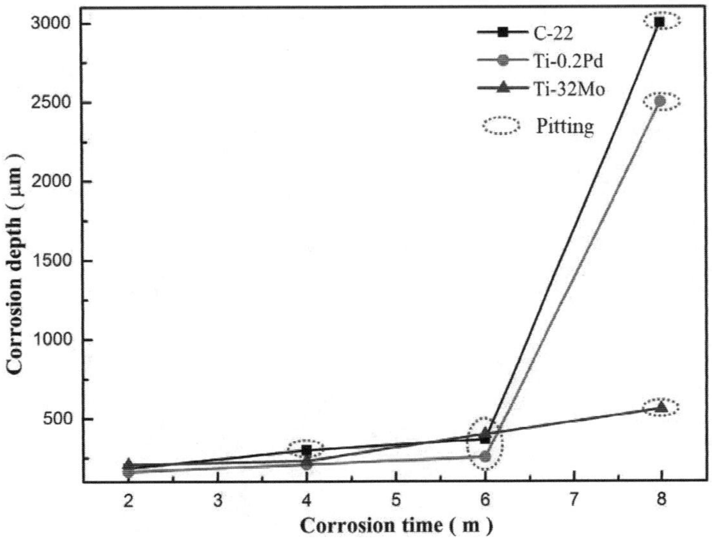

Figure 5. Corrosion depth of specimens after different corrosion time.

3.3. Corrosion mechanism

Typical micrographs of specimens after corrosion are listed in Figure 6. It can be seen that corrosion occurs both at the grain boundary and inside the grain. After corrosion for two months, the grain boundaries appear to be marked and attack in depth is found inside the grain. The grain size of Ti-0.2Pd alloy is the largest, so the area of the grain boundary contacting with the corrosion solution is the least, resulting in the smallest weigh loss rate in the initial period when the grain boundary corrosion dominates.

In the C-22 alloy, elementals Cr, Mo and W are added into the matrix Ni. Cr is needed for corrosion resistance in oxidizing environment, Mo and W are needed for their protection against attack by reducing environments[2] [5] [9]. However, some carbides that form principally at the grain boundaries are generally richer in Cr and Mo than in the matrix; therefore, the matrix immediately adjacent to these carbides are depleted in these elements causing a localized decrease in the corrosion resistance of the alloy. A depletion of Mo causes an increase in the corrosion rate in HCl and a depletion in Cr causes an increase in the corrosion rate in the $NaClO_3$ solution [1]. A thin layer of nickel halide formed by the reaction of Ni with halogen, which is insoluble and has good protective performance, can slow down the corrosion rate [9]. Therefore, pitting corrosion occurs and the weigh loss rate increases slowly in C-22 alloy.

In the Ti-0.2Pd alloy, the addition of Pd increases the exchange current density of Ti and therefore, facilitates passivation in many systems[3]. In oxidizing acid, Ti-0.2Pd alloy is easier to passivate, and the passivation film is hard to break down. However, in reducing solution, the passivation trend decreases and Cl^- will break through the passivation film. The thickness of passivation film is only a few nanometers, so the exposed bulk material [3] inside will be corroded rapidly after contacting with corrosive solution, which is the reason why the weigh loss rate and corrosion depth of Ti-0.2Pd alloy increase sharply after pitting corrosion.

In the Ti-32Mo alloy, Mo can improve the corrosion resistance of Ti in reducing medium due to its high thermodynamic stability, high corrosion potential, and stable bluntness in HCl solution. However, strong oxidants, such as $NaClO_3$, can increase the anode potential, which leads to transpassivation and corrosion of Mo [8]. In the process of corrosion, the products of reaction between Mo and Cl^- form a insoluble and dense protective film [10], which slows down the corrosion rate, that is why the weigh loss rate and corrosion depth increase slowly even if pitting corrosion happens.

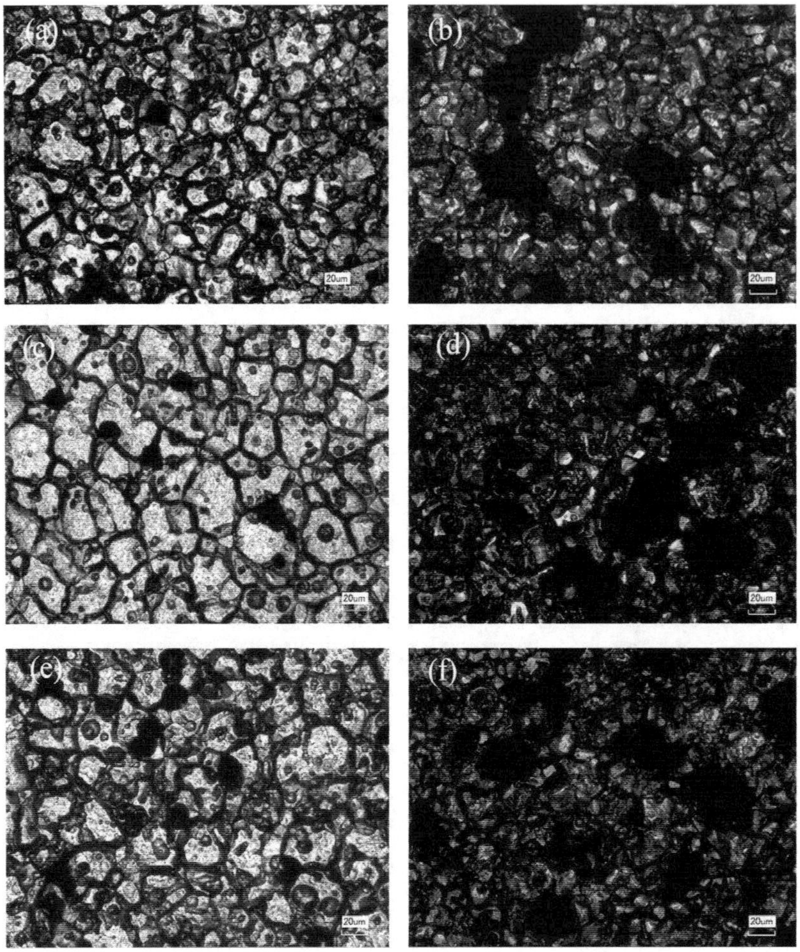

Figure 6. Typical micrographs of specimens after corrosion: a, c and e aremicrographs of C-22, Ti-0.2Pd and Ti-32Mo after corrosion for 2 months, respectively; b, d and f are micrographs of C-22, Ti-0.2Pd and Ti-32Mo after corrosion for 8 months, respectively.

4. Conclusions

Corrosion behavior of C-22, Ti-0.2Pd and Ti-32Mo alloys in the aqueous solution of hydrochloric acid and sodium chlorate was investigated. The main conclusions are the followings:

(1)After corrosion for six months, all the alloys were pitted, and the corrosion rate of Ti-0.2Pd increased dramatically after pitting corrosion.

(2)In order to improve the corrosion resistance in the aqueous solution of hydrochloric acid and sodium chlorate, the content of Cr, Mo and W in C-22 alloy should be adjusted. At the same time, intergranular carbides rich in Cr and Mo need to be suppressed as much as possible.

(3)The passivation state of Ti-0.2Pd alloy in reducing medium is easy to be destroyed. Cl^- will destroy the passivation film, and then corrodes the exposed bulk material inside. Therefore, the Ti-0.2Pd alloy was pitted dramatically when it contacts with strong reducing medium for a long time.

(4)The corrosion rate of Ti-32Mo alloy was the lowest even if pitting occurred. This is because the products of reaction between Mo and Cl^- form an insoluble and dense protective film.

(5)In order to improve the corrosion resistance in the environment where strong oxidant and reductant coexist, it is necessary to pay attention to the titanium alloy with both Pd and Mo added in.

Acknowledgments
This work was supported by the Scientific research project of the Education Department of Sichuan Province (Grant No. 18ZA0013).

References

[1] R. B. Rebak, T. S. E. Summers. R. M. Carranza 2000 Mechanical properties, microstructure and corrosion performance of C-22 alloy aged at 260℃ to 800℃ *Materials Research Society* **608** 109-114
[2] H. M.Tawancy, L. M. Alhems 2018 Correlation Between the State of Atomic Order in Selected (Ni–Mo–Cr)-Based Alloys and Their Corrosion Resistance *Metallography, Microstructure, and Analysis* **7** 746-754
[3] P.L. Cabot, A. Forn, M. Vilarrasa, J.A. Picas, F.J. Gil and J.M. Costa 2000 Effect of heat treatment on the anodic oxidation of Ti-0.2 Pd alloys in chloride solutions *Journal of Applied Electrochemistry* **30** 717-722,.
[4] Amy C. Lloyd, James J. Noël. Stewart McIntyre, David W. Shoesmith 2004 Cr, Mo and W alloying additions in Ni and their effect on passivity *Electrochimica Acta* **49** 3015-3027
[5] X. Wang 1990 Corrosion of Ni-Cr-Mo based corrosion resistant alloy *Instrument materials* **21** 226-237
[6] C. S. Brossia, G. A. Cragnolino 2004 "Effect of palladium on the corrosion behavior of titanium *Corrosion Science* **46** 1693-1711
[7] Y. Ji 1987 Corrosion behavior of Hastelloy C4 in hot brine *Anti corrosion Abstract*
[8] Study on the electrochemical properties of Ti and Ti-Mo alloy *Nonferrous Metal Materials and Engineering* 1-11 1978
[9] C. Li, H. Xu 2011 Analysis and application of corrosion resistance of nickel and nickel based alloy *Henan chemical industry* **28** 18-21
[10] D. Qu, X. Wu. G. Wang 2003 XPS Study on the structure of passivation film of Ti32Mo in concentrated hydrochloric acid solution *Corrosion Science and Protection Technology* **15** 130-133

Ceramic proppants for hydraulic fracturing

A A Biryukova[1], T D Dzhienalyev and A V Panichkin

Satbayev University, Institute of metallurgy and ore beneficiation, Almaty, Kazakhstan

[1]Email: a.biryukova@satbayev.university

Abstract. The purpose of the work is the development of ceramic proppants from serpentinites and red mud. The chemical and mineralogical composition of the raw material was studied. It has been established that the main mineral of serpentinites is serpentine, which is present in the form of fibrous chrysotile and lamellar antigorite. Red mud is represented by sodium aluminosilicates of various hydrated forms, silicon oxide and hematite. Structural-phase transformations of compositions based on serpentinite and red mud were studied in the heat treatment process. The proppant phase composition is represented by forsterite, clinoenstatite, spinel compounds and complex sodium aluminosilicates. Proppants of magnesia-silicate composition based on serpentinite and red mud have been obtained with the following properties: bulk density 1.59-1.61 g/cm^3, fraction of crushed granules with a compressive load of 69 MPa no more than 22%, sphericity and roundness - 0.8, chemical resistance - 98%.

1. Introduction

The hydraulic fracturing is widely used to increase hydrocarbon production in the oil and gas industry in hard-to-develop fields. Hydraulic fracturing is an oil and gas production process that involves injecting water, propping agent and fracturing fluid under high pressure into a rock formation via the well. Propping agents or proppants, fixe cracks in the open state after depressurization [1-4]. Various chemicals are used as a fracturing fluid.

The main functional properties of proppants are crush resistance, chemical resistance and also permeability of proppant pack [2, 3]. Depending on the characteristics of the field, the structure and structure of the oil reservoir, various proppants are required. Proppants vary in composition and properties. The most widely used in hydraulic fracturing were aluminosilicate [4-11] and magnesia-silicate [12-17] proppants. For the production of proppants, both natural raw materials and waste from various industries are used [18-20].

The growth in the extraction and processing of minerals in the world leads to the depletion of natural resources and the accumulation of man-made waste. In this regard, there is an urgent need for the rational use of natural resources and the disposal of man-made waste.

The results of studies on the obtaining of magnesium-silicate ceramic proppants using natural serpentinite and waste from Pavlodar alumina production - red mud are presented. The research results will improve the proppant production efficiency through the use of cheap natural raw materials and alumina production wastes, as well as reducing the energy intensity of the process due to a decrease in the proppant firing temperature.

IOP Conf. Series: Materials Science and Engineering 1040 (2021) 012008 doi:10.1088/1757-899X/1040/1/012008

2. Materials and methods

Chemical analysis of raw materials was carried out in a specialized laboratory using certified methods. X-ray structural analysis of the feedstock and synthesized compositions was carried out on a D8 Advance diffractometer (BRUKER).

Serpentinite and red mud were used as raw materials. The chemical composition of the raw materials is shown in Table 1.

Table 1. Chemical composition of the raw materials (wt%).

Sample/Oxide	SiO_2	Al_2O_3	Fe_2O_3	MgO	CaO	Na_2O	TiO_2	LOI
Serpentinite	34.8	0.9	7.3	39.0	0.8	-	-	17.2
Red mud	17.8	20.0	25.9		0.6	12.2	2.9	19.6

A sample of serpentinite was represented by stone material. Pieces of various colors were present - gray, grayish-mousse, dark-mousse, light with a yellow coating, etc.. Despite the difference in color, all varieties of pieces were serpentinite rock containing the main mineral - serpentine (Figure 1).

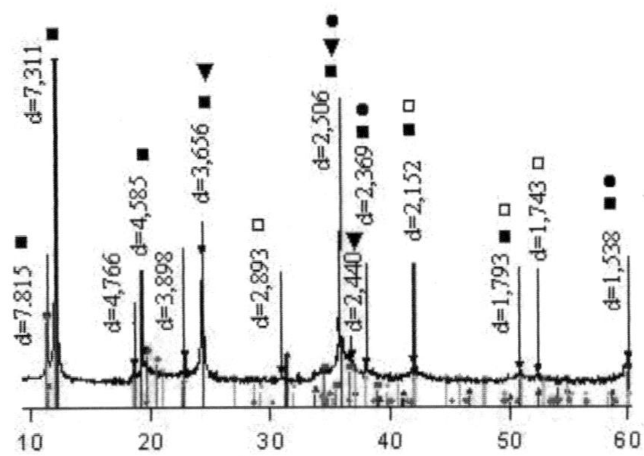

▼ - *hematite,* ■ - *serpentine,* □ - *quartz,* ● - *olivine*

Figure 1. XRD patterns of serpentinite rock.

Red mud is a waste from alumina production containing both alumina and silicon, as well as iron oxides. The phase composition of the red mud sample is shown in Figure 2. The main phases of red mud are hematite, quartz, aluminum silicates $Na_6(AlSiO_4)_6$, $AlNa(SiO_4)$. $Na_{7.6}(Al_6Si_6O_{24})(CO_3)_{0.93}(H_2O)_{2.93}$.

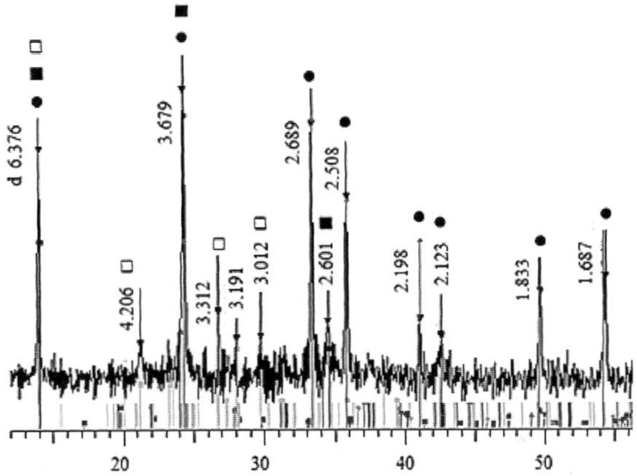

● - *sodium aluminosilicates,* ■ - *hematite,* □ - *quartz*

Figure 2. XRD patterns of red mud.

IOP Conf. Series: Materials Science and Engineering 1040 (2021) 012008 doi:10.1088/1757-899X/1040/1/012008

Two-component mixtures were investigated. Serpentinites were used as the main magnesium and silica-containing component of the charge, and red mud was used as a component with an increased content of iron impurities. Table 2 shows compositions of the studied mixtures.

Table 2. Studied mixtures and their chemical composition.

Mixture	Serpentinite (%)	Red mud (%)	Chemical composition (wt%)				
			SiO_2	Al_2O_3	Fe_2O_3	MgO	Others
1	100	-	42.0	1.08	8.85	47.1	0.97
2	90	10	40.04	3.43	11.25	42.5	2.78
3	80	20	38.06	5.76	14.1	37.38	4.66
4	70	30	36.12	8.25	15.93	32.93	6.77
5	60	40	34.16	10.65	18.31	28.26	13.28

Presumably, silicon oxides should contribute to the formation of magnesium metasilicate during the firing of the ceramic composition, aluminum and iron oxides to the formation of spinel phases, which can increase the strength properties of proppants. The assimilation of sodium-containing compounds will contribute to the formation of low-temperature eutectics, which should reduce the temperature of sintering proppant firing. With a decrease in the amount of serpentinite in the mixture, the content of silicon and magnesium oxides decreases, while the content of aluminum, iron and sodium oxides increases. Depending on the composition of the charge during the firing process, certain phases will be formed, and their ratio will change.

3. Experimental part and results

Earlier, in [21], the efficiency of preliminary thermal activation at 1000 °C of Kazakhstan serpentinite raw materials for the production of ceramic materials was established. The materials were preliminarily subjected to heat treatment at 900-1000 °C in order to remove chemically bound water. In order to assess the strength properties and phase composition of the compositions based on a mixture of serpentinite and red mud, cylinder samples with a diameter of 16 mm were made by pressing. The prepared samples were fired in the range of 1200-1350 °C in an electric furnace with subsequent determination of properties (Table 3). Samples of serpentinite used as a standard of comparison.

Table 3. Properties of ceramics.

Sample	Firing temp. (°C)	Water absorption (%)	Apparent density (g/cm^3)	Compressive strength (MPa)
1	1350	7.5	2.72	65
2	1300	4.5	2.70	71
3	1240	2.8	2.67	75
4	1230	2.1	2.68	78
5	1200	0.9	2.68	69

The test results showed that samples of composition 1 (standard) have the minimum strength. X-ray phase analysis (Table 4) of the products of firing of composition 1 revealed the presence of the main phase of forsterite, which has anisotropic properties, poorly sintering, which prevents the compaction and hardening of the ceramic structure.

The introduction of red mud in the mixture leads to an increase in the content of iron, aluminum and sodium oxides. The presence of sodium oxides (Figure 3) in the composition contributes to the formation of low-temperature eutectics, activating the process of sintering of the mass due to the participation of the melt, which contributes to the compaction and hardening of the ceramic structure (compositions 2-5). The phase composition of the remaining samples containing red mud is similar to sample 3 with a difference in a phase ratios.

Table 4. Phase composition of ceramics.

Samples	Phases
1	Forsterite, ferroan $(Fe_{0.12}Mg_{0.88})_2(SiO_4)$, Magnesium Iron Silicate $(Mg_{0.74}Fe_{0.26})(Si_2O_6)$, Magnesioferrite aluminian $(MgAl_{0.74}Fe_{1.26})O_4$.
3	Forsterite $(Mg_{1.81}Fe_{0.19})(SiO_4)$, Nepheline, $Na_{6.65}Al_{6.24}Si_{9.76}O_{32}$, Jadeite $NaAl(SiO_3)_2$, Spinel $Mg_{1.01}Fe_{1.77}Al_{0.22}O_4$, Maghemite, syn (Fe_2O_3), Clinoenstatite ferroan $(Mg_{0.54}Fe_{0.46})SiO_3$.

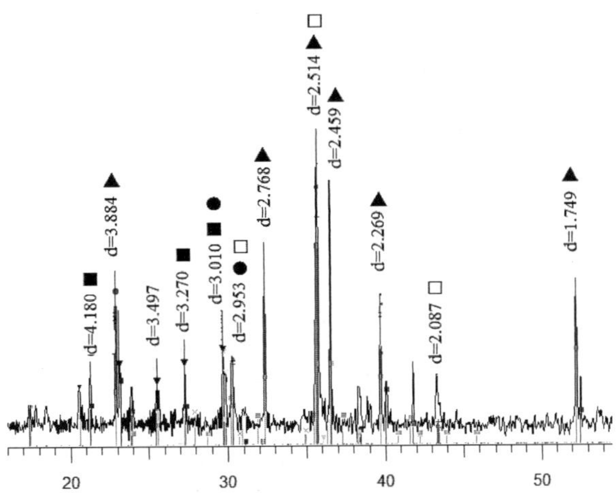

▲ - *forsterite,* ■ - *nepheline,* □ - *spinel,* ● - *clinoenstatite*

Figure 3. XRD patterns of the sample 3.

This has a positive effect on lowering the temperature of sintering proppant firing. An increase in the content of red mud (more than 40%) leads to a significant formation of low-temperature phases. An increase in the amount of melt during the firing process promotes wetting and sintering of solid particles. However, upon cooling of the samples, the melt crystallizes in the form of glass mass and the strength properties of the samples decrease.

Reducing the content of red mud in the composition of the charge of less than 20% does not provide a sufficient content of sodium oxides, necessary to reduce the firing temperature of proppants.

The presence of aluminum and iron oxides in the charge facilitates the formation of defective solid solutions of interstitial and (or) iron substitution in the lattice of metaserpentin and magnesium orthosilicate nuclei, as well as in the formation of sintering complex magnesia spinels that activate sintering. This effect from the presence of Fe2O3 in the composition of the charge is possible due to the proximity of ionic radius of Al3+ (0.057 nm) and of Fe3+ (0.067 nm). The presence of spinel phases in the structure of ceramics increases the strength of samples and chemical resistance to aggressive solutions.

The research results showed that all obtained ceramic samples by phase composition belong to magnesia-silicate. The main magnesia-silicate phases of all compositions are forsterite and clinoenstatite, the content of which depends on the composition of the charge. The formation of solid solutions of magnesia and iron silicates has also been established. Samples of magnesium-silicate ceramics obtained from a mixture containing from 10 to 30% red mud have maximum strength.

Proppants were obtained from compositions containing 10-30% red mud. Proppants were prepared as follows. The raw materials were subjected to grinding to a fraction of 0.063 mm, followed by heat treatment of the mixture at 900-1000 °C in order to: remove constitutional moisture; decomposition of the initial mineral serpentine with the formation of metastable metaserpentine and magnesium orthosilicates with a defective structure. This will activate the sintering process and densification of the ceramic structure in sintering firing.

Structural-phase transformations during heat treatment of serpentinite rock with the formation of forsterite through the metastable phase of metaserpentine in the range of 200–1000 ° C were established earlier in [21].

Heat-treated fine-ground powders of raw materials (fraction less than 0.063 mm) were loaded into a granulator-mixer Eirich EL1 to obtain proppant granules. The mixing of the mass took place for 30 seconds, at a maximum speed of rotation of the swirler. Then introduced the binder component of a 1.5% solution of methylcellulose in an amount of 20% of the total mass of the charge. The mass was mixed for 30 seconds, then the rest of the ligament was gradually introduced. After supplying the entire volume of the binder component, the mass was mixed for another 1 minute to evenly moisten and compact the mass.

The swirler speed was reduced to 700 rpm at the time of the appearance of small granules (0.1-0.3 mm). At this time, crushed powder for dusting in an amount of 10-15% of the charge necessary to obtain granules of a given size was additionally uniformly introduced into the granulator. After 1-3 minutes after dusting, the raw granules were unloaded, dried under natural conditions during the day, then at a temperature of 105-115 °C for 40-60 minutes to a residual moisture content of 1-3 mass. %. The proppants were roasted at a sintering temperature, depending on the composition of the charge, in the range of 1,200-1,250 °C for 1 hour.

After cooling, the calcined proppant granules were fractionated to isolate the 16/20 fraction, which was tested to determine the properties in accordance with GOST R 54571-2011 (KZ). Results of testing are shown in Table 5.

Table 5. Properties of proppants.

Mixture	Firing temp. (°C)	Bulk density (g/cm^3)	Crush loss under 69 MPa (%)
1	1350	1.59	23.9
2	1300	1.61	19.4
3	1240	1.60	21.8
4	1230	1.59	21.5
5	1200	1.59	22.1

The test results showed that the obtained ceramic proppant granules are proppants of medium density with a bulk density of 1.59-1.61 g/cm^3. The crush resistance of the 16/20 fraction at a pressure of 69 MPa is 19-23%. Proppant granules were also tested for corrosion resistance to a solution of 10% hydrochloric acid. The sphericity and roundness of the obtained proppants is 0.8, and the chemical resistance to hydrochloric acid is 98%. Tested proppant samples of all compositions according to their technical properties comply with the requirements of GOST R 54571-2011 (KZ) for magnesium-quartz proppants.

4. Conclusions

Studies on the development of ceramic proppant compositions based on serpentinites of the Kempirsai region and red mud of Pavlodar alumina production have been carried out.

1. The chemical and mineralogical composition of serpentinite and red mud has been studied. It has been established that the main mineral of overburden is serpentine, which is present in the form of

fibrous chrysotile and plate antigorite, in the form of impurities there are iron oxides and hydroxides, carbonates, quartz, hydromagnesite, talc. The main components of red mud are oxides of aluminum, iron, silicon and sodium.

2. It was found that during the heat treatment of compositions based on serpentinite and red mud, structural-phase changes of the initial minerals occur with the formation on their basis of the main phases of magnesia-silicate ceramics: forsterite and clinoenstatite, the content of which depends on the composition of the charge. The formation of solid solutions of magnesia and iron silicates and impurity sodium aluminosilicates has also been established.

3. Obtained proppants with a bulk density of 1.6 g/cm3; with crush resistance for granules 16/20 no more than 25%; sphericity and roundness - 0.8; chemical resistance to hydrochloric acid - 98%.

Acknowledgments
This research was financially supported by the Science Committee of the Ministry of Education and Science of the Republic of Kazakhstan (Grant No. AP05131248).

References
[1] Li Q, Xing H, Liu J, Liu X 2015 *Petroleum* **1** 8
[2] Kuznetsov D, Kulagina T, Malakhov L. Hydraulic fracturing. Petroleum Learning Centre, Tomsk, 2015
[3] Vasilev V, Verisokin A 2013 *Vestnik PNIPU. Geology (In Russian)* **6** 102
[4] Khasanova A, Almukhametova E, Gabdrakhmanov N 2016 *Problemy sbora, podgotovki i transporta nefti i nefteprodutov (In Russian)* **1** 7
[5] Liang F, Sayed M, Al-Muntasheri G, Chang F, Li L 2016 *Petroleum* **2** 26
[6] Campos V, Sansone E, Silva G 2018 *Cerâmica* **64** 219
[7] Ma X, Tian Y, Zhou Y, Wang K, Chai Y 2016 *Material Letters* **180** 127
[8] Xianfeng Z, Qinchang J 2017 China Patent CN106967409A
[9] Mozherin V, Novikov A, Sakulin V 2017 Russia Patent 2619603
[10] Zhao J, Liu Z, Li Y 2015 *Materials Letters* **152** 72
[11] Alekseev V, Shmotiev S, Pliner S 2010 Russia Patent 2381202
[12] Alekseev V, Dykov A, Antonov A 2012 Russia Patent 2452759
[13] Biryukova A, Dzhienalyev T, Tikhonova T 2017 *Refractories and Industrial Ceramics* **58** 269
[14] Qiang R, Yuhan R, Huanhuan Li, Xiulan Wu 2019 *Materials Chemistry and Physics* **230** 355
[15] Shmotiev S, Pliner S, Sychev V 2010 Russia Patent 2646910
[16] Plotnikov V, Pliner A, Pejchev V 2014 Russia Patent 2521989
[17] Shmotiev S 2015 Russia Patent 2563853
[18] Kramar L, Chernykh T, Orlov A, Prokofieva V 2011 *Dry mixes (In Russian)* **2** 14-17
[19] Rusinov P, Balashov A 2017 Russia Patent 2615563
[20] Plotnikov V, Rozhkov E, Pejchev V 2014 US Patent 8,785,356
[21] Biruykova A., Dzhienalyev T., Boronina A 2018 *News of the Academy of Sciences of the Republic of Kazakhstan* **431** 114-119

Research on the wear and life of copper tungsten alloy contacts of high voltage SF_6 circuit breakers

Chao Xu[1,4], Hai Huang[1], Ying Wang[2] and Xingcai Shi[3]

[1]State Grid Heilongjiang Electric Power Company Limited Electric Power Research Institute, 7 Xiangjiang River Road, Harbin, China;
[2]Economic and Technological Research Institute of State Grid Heilongjiang Electric Power Co., Ltd., Harbin, China;
[3]State Grid Liaoning Electric Power Supply Co.,Ltd Skills Training Center, Jinzhou,China

[4]Email: 704777887@qq.com

Abstract. Three phase short circuit and two phase short circuit grounding fault of power system have a significant influence on the electrical life of circuit breaker contact. In this paper, the mathematical model of each component is established by Power System Analysis Software Package (PSASP), and the simulation model is built to simulate the dynamic state of the circuit breaker after different short-circuit faults. The simulated result confirms that contact electrical wear quantity related to the type of short circuit fault and the position of short circuit point. The size of short circuit current will directly impact electrical wear quantity and electrical life of circuit breaker contact.

1. Introduction

In power system operation, high voltage circuit breaker is switching current device. Its electrical life is generally shorter than the overall life of high voltage circuit breakers. The electrical life of the high voltage circuit breaker is decided by the contact and the arc extinguishing chamber, and the maintenance and replacement are relatively easy. On the one hand, through the on-line monitoring of the electrical life of the high voltage circuit breaker, the service life of the high voltage circuit breaker can be extended when its performance decline in time to repair or replace. On the other hand, it can also avoid blind maintenance to extend the use of equipment. With the continuous development of condition based maintenance, wear of high voltage circuit breaker contact analysis is very necessary, the production and manufacture of high voltage circuit breaker also has certain reference[1-3].

2. Simulation model of high voltage circuit breaker based on PSASP

2.1. Power System Simulation Model

The network structure of power system simulation model is illustrated in Figure 1, including generator system (power system stability system (PSS), excitation regulation system, speed control system) and power network.

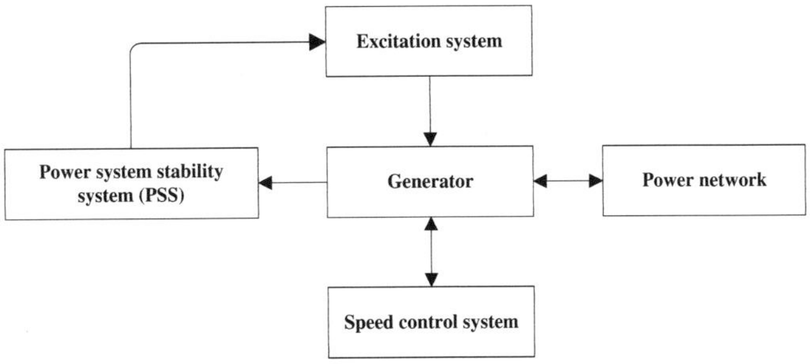

Figure 1. Power system simulation model.

2.2. Mathematical model and PSASP simulation model

$$T_{d0}^{'}\frac{dE_q^{'}}{dt}=E_{fd}-[E_q^{'}+(x_d-x_d^{'})I_d+(K_G-1)E_q^{'}] \quad (1)$$

$$T_{q0}^{'}\frac{dE_d^{'}}{dt}=-E_d^{'}+(x_d-x_q^{'})I_q \quad (2)$$

$$T_J\frac{d\omega}{dt}=\frac{P_T}{\omega}-(\psi_d I_q-\psi_q I_d)-D(\omega-\varpi_0)2\pi f_0 \quad (3)$$

$$\frac{d\delta}{dt}=(\omega-1)2\pi f_0 \quad (4)$$

x_d, generator d-axis synchronous reactance; $x_d^{'}$, generator d-axis transient reactance; K_G, saturation coefficient; D, equivalent damping coefficient; $x_q^{'}$, generator q-axis transient reactance; ψ_d and ψ_q, stator winding flux; T_{d0}', the open circuit time constant of generator rotor d-axis excitation winding stator; T_{q0}', the open circuit time constant of generator rotor q-axis damper g winding stator; T_J, generator rotor inertia time constant $T_J=2H$; I_d, generator rotor d-axis winding current; I_q, generator rotor q-axis winding current; E_{fd}, excitation voltage; $E_q{}'$, transient potential saturation value.

2.3. Power system stability model (PSS)

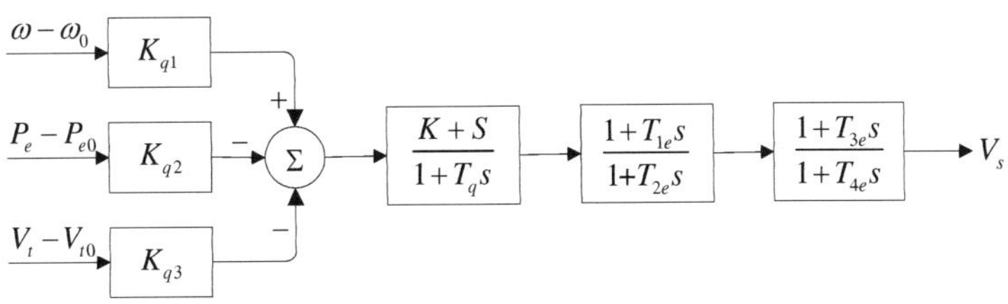

Figure 2. Power system stability model (PSS).

K_{q1}, rotating speed error amplification factor; K_{q2}, electromagnetic power error amplification factor; K_{q3}, voltage deviation amplification factor; T_q, blocking link time constant; T_{1e}, T_{2e}, T_{3e}, T_{4e} phase shift link time constant; K, change link $\frac{K+S}{1+T_q S}$ coefficient of type; K=0, the straight link (inertial differential link); K=1, the phase shift link; PSS output, these amounts are shown in the Figure 2.

2.4. *Excitation system model*

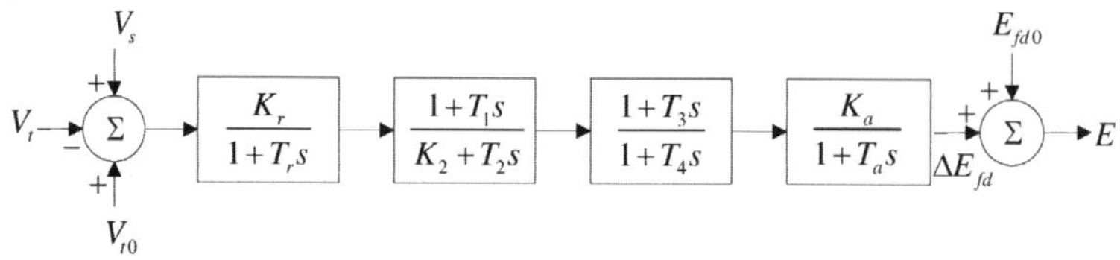

Figure 3. Power system stability model (PSS).

V_t, machine terminal voltage; V_{t0}, the initial voltage of the machine; K_r, the amplification factor of the measuring link; K_a, the amplification factor of the amplifying link; K_2, transformation link type parameter; T_r, measurement link time constant; T_a, amplification link time constant; T_1, T_2, T_3, T_4 intermediate link time constant, these amounts are shown in the Figure 3.

2.5. *Speed control system model*

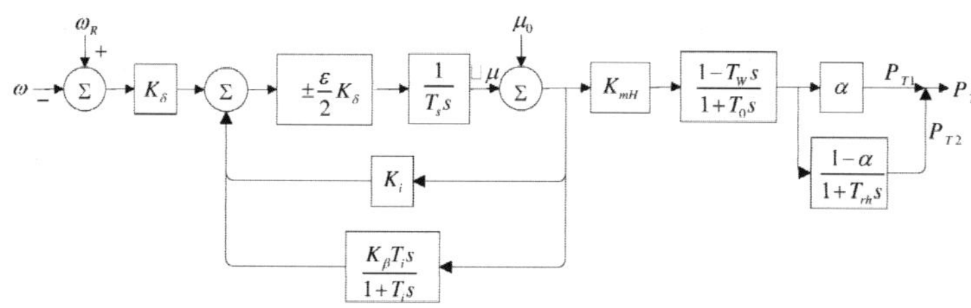

Figure 4. Speed control system model.

K_σ, measurement link amplification factor; K_i, negative feedback amplification factor; K_β, negative feedback amplification factor; ε governor dead zone; α turbine over heat coefficient; T_i, soft feedback time constant of turbine; T_s, time constant of servo mechanism; T_0, the time constant of the steam volume; T_W, water hammer effect time constant; T_{rh}, turbine intermediate superheating time constant; μ_0, guide vane (valve) initial opening; K_{mH} equivalent to the ratio of the generator's rated power to the system reference capacity in fact, $K_{mH} = \frac{P_H}{S_B}$, P_H System reference capacity, S_B rated power of generator, these amounts are shown in the Figure 4.

3. Study on calculation method for electrical wear and electrical life of high voltage circuit breaker contact

3.1. *Main calculation method for electric wear of circuit breaker contact*

The main characteristic parameter of electric life of circuit breaker is electric wear, this paper discusses the circuit breaker electrical wear refers to the wear amount of circuit breaker contact, not including the wear of arc extinguishing chamber and arc extinguishing media. At present, the main methods to calculate the electrical life of circuit breaker contacts are:

3.1.1. *The cumulative breaking current method or the cumulative arc energy method*

$$\frac{d\delta}{dt} = (\omega - 1)2\pi f_0 \quad (5)$$

$$Q = \sum_{L}^{N} I_{ikd}^{2} t_i \tag{6}$$

Q, the total electrical wear of circuit breaker; N, rated breaking current of circuit breakers to allow the number of open and broken; I_{ikd}, single breaking current value of circuit breaker; t_i, single break time of circuit breaker.

This method is widely used in the early detection of circuit breaker, although it only needs to measure the current effective value, but its sensitivity is not high.

3.1.2. The method of weighted cumulative current

$$Q = \sum_{L}^{N} (I_{bkd})^{\alpha} \ (1 < \alpha < 2) \tag{7}$$

$$Q_g = N \times (I_{bikd})^{\alpha} \quad (1 < \alpha < 2) \tag{8}$$

Q_g, absolute total electrical wear of circuit breakers; I_{bikd}, single break current of circuit breaker considering the material and the arc extinguishing medium; I_{bkd}, single rated breaking current of circuit breaker considering the material and the arc extinguishing medium; α, a constant, value between 1~2, generally take 2, which is related to the circuit breaker contact material.

This method is widely used, which considers the influence of the material of the circuit breaker, the arc extinguishing medium and other factors on the electrical wear of the circuit breaker contact. This method is relatively simple, but the disadvantage can only be found out of the electrical wear.

3.1.3. The arc time weighted evaluation method

$$Q_g = K_{\omega} K_g \sum_{L}^{N} \int_{0}^{\alpha i} [ibi(t)]^{\beta} \, dt \tag{9}$$

K_{ω}, constant value, related to the principle of arc extinguishing in different media, The cooling conditions, contact switching speed and material of contact; K_g, ratio coefficient; β, a constant, value between 1~2, generally take 2, which is related to the circuit breaker contact material.

This method can accurately judge the electrical life of circuit breakers, mainly due to respectively cumulative metering the three-phase contact electrical wear condition and the arc burning time is considered. Category was used to calculate the single disconnect current wear amount. But in the course of using, the need to install the device to measure the arc time is cumbersome.

3.1.4. Relative electrical wear and relative electrical life of the contacts. Considering the circuit breaker in the actual operation process, when the breaking current value difference is large, the difference of AC and DC components of the breaking current is also great and the extent of burning contact is different. But even if the same size of the current, the contact erosion degree is not the same due to the randomness of arc quenching process. Using the method of cumulative breaking current to calculate the amount of contact burning is not careful. In determining the electrical life of the circuit breaker, the cumulative current weighted method and the weighted method of time weighted all considered that electrical wear is related not only to the arc energy, also with the contact points off speed and arc quenching medium factors. This improved than before. But from the point of view of engineering practice and accuracy of judgment, it is feasible to calculate the relative electrical wear and relative electrical life of the contact.

3.2. Method for calibrating electrical life of circuit breaker contact

Contact by repeatedly opening and closing, the contact surface will gradually occur deformation, welding, surface potholes and other phenomenon, this phenomenon called the abrasion of the contact. The wear of the contact is mainly electrical wear, the core of which is the metal volatilization of contact material, transfer, chemical corrosion and the net loss of contact material. The energy of arc breaking current and arcing time mainly determines the wear contact, Fusing or pneumatolytic contact surfaces will cause the loss of contact by the fluid medium washing or spattering. In the course of each opening and closing of breaker, arc combustion can not be the same, but from a statistical point of view, it still follows a certain law. Considering the actual operation of the circuit breaker, the average time of the circuit breaker tends to be stable when the number of open and close circuit breakers is more frequent, from this point of view, the breaking current can be used as a reference to determine the cumulative effect of circuit breaker, and the randomness of arcing time is not considered.

The definition of electrical wear of rated short circuit breaking current is M_a, the allowable number of breaks is N_a ,from the cumulative effect and statistical analysis, it can be concluded that the total amount of the circuit breaker wear is $M_a N_a$. Setting the total wear of a new breaker contact is 100%, and its relative electrical life is 100%, the relative wear of the corresponding single rated short circuit breaking current is $1/N_a$, it will be defined as Q_a. According to the $N-I_c$ curve of different type circuit breakers, the allowable breaking times N corresponding all the different size of the breaking current value I_c can be calculated. The absolute amount of electrical wear about a single break is M, the relative wear is 1/N, which is defined as Q.

Based on the above definition, the relative wear amount of circuit breaker contact in one break:

$$Q_a=1/N_a \tag{10}$$

$$Q=1/N \tag{11}$$

Since the total amount of absolute electric contact wear is a constant value, the formula as following:

$$Q_a N_a=QN \tag{12}$$

$$MN=M_a N_a \tag{13}$$

$$Q/Q_a=N_a/N=M/M_a \tag{14}$$

$$Q=Q_a(M/M_a)=M/M_a N_a \tag{15}$$

Therefore, the relative electrical life of the circuit breaker is:

$$L=L_N-\sum Q \tag{16}$$

In the above formula, the L_N is defined as the starting value of the relative electrical life of the circuit breaker contact, which is a number of less than or equal to 1, and the operating conditions of the circuit breaker determine its value. Just after the overhaul, replacement or newly installed circuit breaker, L_N can take 1.

3.3. Calculation method of residual electrical life of SF_6 circuit breaker

Known a SF_6 circuit breaker allows the number of consecutive break current I_{sn} is N, then contact burning amount corresponding to a single break is 1/N. In any breaking current $I_s<0.35I_{sn}$, are discussed in this paper from the front, with a break burning I_{sn} the equivalent I_s breaking times for N_{s1}.

$$N_{s1}=1.83(0.35\times I_{sn}/I_s)^3\times(0.5\times I_{sn}/I_{sn})^{-1.7}=5.95(0.35\times I_{sn}/I_s)^3 \tag{17}$$

Corresponding to an breaking I_s, the contact burning amount should be:

$$Q_{s1}=\frac{1}{N_{s1}}\times\frac{1}{N}=\frac{1}{5.95}\times(0.35I_{sn}/I_s)^{-3}\times\frac{1}{N} \tag{18}$$

Corresponding to any breaking $I_s \geq 0.35I_{sn}$, the contact burning amount should be:

$$Q_{s2}=\frac{1}{N_{s2}}\times\frac{1}{N}=\frac{1}{3.25}\times(0.5I_{sn}/I_s)^{-1.7}\times\frac{1}{N} \tag{19}$$

Make new circuit breaker relative electrical life is 1, running after a period of time the cumulative relative loss is $\sum Q_{si}$, relative surplus electrical life:

$$L = L_N - \sum Q_{si} \tag{20}$$

In the above formula, L_N is the initial value of the relative electrical life of the circuit breaker, and the average value is 1. $\sum Q_{si}$ is the contact wear was calculated by the former. $L > 0$, can consider the circuit breaker status is normal, $L \leq 0$ cannot continue to run, need maintenance.

4. Simulation calculation analysis of SF_6 circuit breaker contact relative wear amount

Simulation model selection of high voltage SF6 circuit breaker types for LW-252/4000-40, rated voltage of 220kV, 4000A of rated current, rated short-circuit breaking current 40kA, rated short circuit currents of 100kA, rated breaking number 20, inherent closing time 100ms, split time 33ms. Take the parameters $\alpha = 3$, $\beta = 1.7$, and when $I_s < 0.15I_{sn}$ there is no wear of Q_m.

$$Q_m=\frac{1}{5.95N(0.35I_{sn}/I_s)^{\alpha}} \tag{21}$$

When: $0.15I_{sn} \leq I_s < 0.35I_{sn}$

$$Q_m=\frac{1}{3.25N(0.5I_{sn}/I_s)^{\beta}} \tag{22}$$

When: $I_s \geq 0.35I_{sn}$

In the simulation of different short circuit (three-phase short-circuit and two-phase short circuit) fault, by changing the length of the line at both ends of the short circuit point to change the location of the fault point. Short circuit current values under different fault types are shown in Figure 5 and Figure 6.

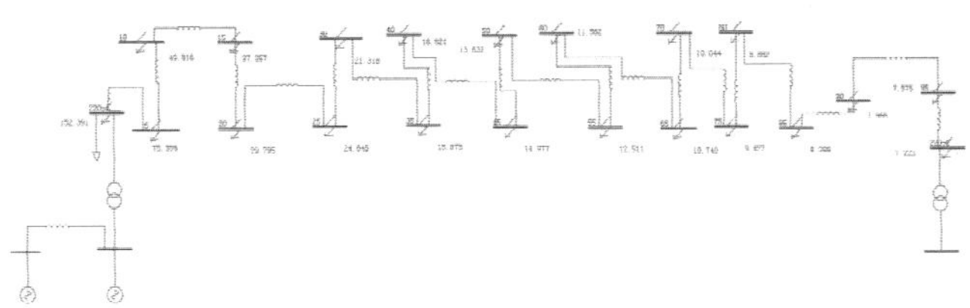

Figure 5. Fault current value of different fault point in three phase short circuit fault

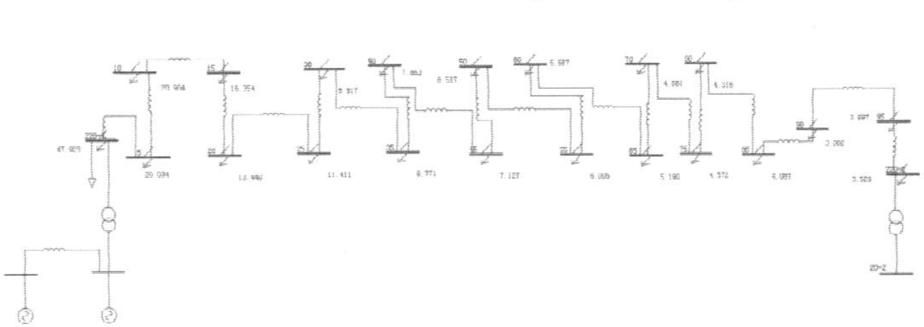

Figure 6. Fault current value of different fault point in two phase short circuit fault.

Comparison of relative electric wear and breaking frequency of circuit breaker under different fault conditions is shown in Table 1.

Table 1. Comparison of relative electric wear and breaking times of circuit breaker under different fault condition.

Fault type	Fault point distance DL(%)	Fault current I_{sn} (kA)	I_{sn}/I_s (%)	The relative amount of wear Q_m	Breaking frequency N
Three phase short circuit fault	0%	40	100%	0.05	20
	5%	19.8	49.5%	0.015123	66
	10%	13.1	32.75%	0.006885	145
	15%	9.8	24.5%	0.002882	347
	20%	7.8	19.5%	0.001453	688
	25%	6.5	16.25%	0.000841	1189
	≥30%	≤5.6kA	≤14%	/	No wear
Two phase short circuit fault	0%	12.6	31.5%	0.006126	163
	5%	7.6	19%	0.001344	744
	≥10%	≤5.5	≤13.75%	/	No wear

5. Conclusion

The first paragraph after a heading is not indented (Bodytext style).1)Contact wear quantity was related with the short circuit type and the position of the short circuit point, and the size of short circuit current directly affect the wear amount of contact.

Relative wear volume becomes larger when the current becomes larger, and at the same point the three phase short circuit fault is larger than that of two phase short circuit fault, thus the relative wear increases. Due to the non periodic component of short-circuit current, the total effective value of the short circuit current will increase, and the arc energy will be more than no the non periodic components. According to the relation between energy and burning, DC component will also affect the degree of contact wear, so in the actual calculation breaking times should be left a certain margin.

Acknowledgements

This work was supported by Science and Technology Project of SGCC (Research on Low-temperature Resistance Characteristics and Evaluation Technology of New Sensors for Power Transmission and Transformation).

References

[1] Hu Wenping, Yin Xianggen, Zhang Zhe, Wu Jieyu, and Shao Dejun 2003 Electrical endurance simulation of high voltage circuit breaker contact based on matlab *Proceedings of the EPSA* **15** 9-12

[2] J Tepper, M Seeger, T Votteler, et al. 2006 Investigation on Erosion of Cu/w Contacts in High-Voltage Circuit Breakers*[J]. IEEE Transaction on Components and Packaging Technologies* **29(3)** 658-665

[3] John J Shea 2008 High Current AC Break Arc Contact Erosion[C]//The 54th IEEE Holm Conference on Electrical Contacts, Orlando 22-46

Synthesis and crystal structure of the one-dimensional chain triphenyltin compound bridged by 5-chloro-6-hydroxynicotinic acid

Zhongjun Gao[1], **Haifeng Zhang and Jingquan Sha**

Department of Chemistry and Chemical Engineering, Jining University, Qufu, 273155, Shandong, P.R. China

[1]Email: zhongjungao@sina.com

Abstract. Reaction of 5-chloro-6-hydroxynicotinic acid with Ph_3SnCl and sodium ethoxide in 1:1:1 stoichiometry yielded triphenyltin compound. This compound has been characterized by single crystal X-ray diffraction, elemental analysis, IR and ^1H NMR spectroscopy. In the molecular structure of the title compound, the tin atoms are five-coordinated in a distorted trigonal bipyramidal geometry. The resulting structure is 1D linear polymer through an interaction between the O atoms of phenolic hydroxide and tin atoms of an adjacent molcule. The compound exhibit high antitumor activity.

1. Introduction

Organotin compounds are widely used as fungicides, biocides and in industry as homogeneous catalysts[1-3]. Recently, pharmaceutical properties of organotin esters of carboxylic acids have been investigated for their antitumour activity[4-7]. Many triphenyltin compounds with ketocarboxylic and hydroxyarylcarboxylic acids and many other N and O containing derivatives of carboxyllic acids display high antitumor activities. However triphenyltin ester 5-chloro-6-hydroxynicotinic acid were not investigated. In order to continue exploring relationships between the biological activity and structure, We choose another ligand: 5-chloro-6-hydroxynicotinic acid. This carboxylic acid is interesting because of its potential multidentate coordinate possibilites[8].

2. Experimental

2.1. Materials and measurements

Triphenyltin chloride and 5-Cl-6-OH-$C_5H_2NCO_2H$ were purchased from Aldrich. Analytical grade solvents were dried before use. The melting points were obtained with Kofler micromelting point apparatus and are uncorrected. Infrared-spectra (KBr pellets) were recorded on a Nicolet-460 spectrophotometer. ^1H NMR spectra were taken on a Varian Mercury Plus 400 spectrometer. The chemical shifts were given in ppm in $CDCl_3$ solvent. Carbon, hydrogen and nitrogen analyses were performed on a PE-2400II apparatus.

2.2. Synthesis of the title compound

The reaction was carried out under N_2 by use of standard Schlenk techniques. The 5-Cl-6-OH-$C_5H_2NCO_2H$ (0.521 g, 3 mmol) was added to the solution of dry C_2H_5OH (50mL) together with C_2H_5ONa (0.214 g, 3 mmol), and the mixture was stirred for 0.5 h. Ph_3SnCl (1.165 g, 3 mmol) was then added to the mixture, and the reaction was allowed to continue for 16 h at 45°C. After cooling down to 20°C and then filtered. The filtrate was gradually removed by evaporation under vacuum until solid product was obtained. Colorless block-like crystal was recrystallized from 1-propyl alcohol. Yield 68 %, m.p. 216-217°C. ^{1}H NMR ($CDCl_3$, 400 MHz): δ 8.88(1H, d, 2-pyridine-H),7.99(1H, d, 4-pyridine-H), 6.78-7.37(15H, m, Ph-H), 0.91-3.51(7H, m, $CH_2CH_2CH_3$). IR (KBr)(cm^{-1}): v_{as}(COO), 1643; v_s(COO), 1355; v(Sn-C), 569; v(Sn-O), 458. Anal. Calcd for $C_{27}H_{25}ClNO_4Sn$: C 55.75, H 4.33, N 2.41. found: C 57.77, H 4.36, N 2.42.

2.3. Determination of the crystal structure

A single crystal having approximate dimensions of 0.23mm × 0.16mm × 0.12mm was mounted in a glass capillary. All measurements were made on a Bruker Smart-II diffractometer equipped with a graphite-monochromated Mo-Kα (λ=0.71073 Å) radiation at 298(2) K by using the φ-ω scan technique. For crystal of the compound, a total of 6290 reflections were collected in the range of 1.96<θ<25.01° and 3672 (R_{int}=0.0269) were independent. The structure was solved by direct methods and difference Fourier map with SHELXL-97 program[9], and refined by full-matrix least-squares techniques on F^2. All non-hydrogen atoms were refined anisotropically, and the hydrogen atoms were located at the calculated positions and refined isotropically. Crystallographic data are listed inTable 1. The selected bond lengths and angles for compound 1 are given in Table 2.

Table 1. Crystallographic data for the compound.

Molecular formula	$C_{27}H_{25}ClNO_4Sn$	Z	4
Formula weight	581.62	Theta range/°	1.96-25.01
Crystal system	Monoclinic	μ/mm^{-1}	1.170
Space group	Cc	Dc / g·cm^{-3}	1.503
a/Å	16.436(6)	F(000)	1172
b/Å	14.562(5)	Goodness-of-fit (F^2)	1.005
c /Å	11.466(4)	R1, wR2 [I>2σ(I)]	0.0461, 0.1235
β / (°)	115.181(4)	R1, wR2 (All data)	0.0506, 0.1302
V ($Å^3$)	2483.9(15)	(Δρ)max, (Δρ)min / (e·$Å^{-3}$)	0.113, -0.985

Table 2. Selected Bond Lengths [Å] and Angles [°] for the compound.

Bond lengths lengths[Å]		Bond angles angles[°]	
Sn1-C19	2.103(6)	C19-Sn1-C7	116.8(3)
Sn1-C7	2.130(4)	C19-Sn1-C13	122.3(4)
Sn1-C13	2.133(9)	C7-Sn1-C13	116.8(4)
Sn1-O1	2.138(6)	C19-Sn1-O1	104.6(3)
Sn1-O3#1	2.662(8)	C7-Sn1-O1	89.4(3)
O1-C1	1.298(5)	C13-Sn1-O1	95.7(3)
O2-C1	1.231(9)	C19-Sn1-O3#1	83.9(3)
C1-C3	1.449(11)	C7-Sn1-O3#1	81.9(3)
C2-C3	1.378(11)	C13-Sn1-O3#1	83.9(3)
O3-C6	1.230(9)	O1-Sn1-O3#1	169.9(2)

Symmetry transformations used to generate equivalent atoms:
#1: x-1/2, y+1/2, z.

3. Results and discussion

3.1. IR Spectra

The assignment of IR bands of the title compound has been determined by comparison with the other two compounds: 5-Cl-6-OH-$C_5H_2NCO_2H$ and Ph_3SnCl. The broad absorption at 3000-2500 cm^{-1} in

IOP Conf. Series: Materials Science and Engineering 1040 (2021) 012010 doi:10.1088/1757-899X/1040/1/012010

free acid due to –OH group, which is absent from the spectra of the coordination of carboxylate group, showing the deprotonation of the -COOH group during the reactions. Asymmetric stretching vibration peak of carboxyl (ν_{as}(COO)) is located at 1643 cm^{-1} and the corresponding symmetric one (ν_s(COO)) at 1355 cm^{-1}. The magnitude of Δv [ν_{as}(COO) - ν_s(COO)] is above 288 cm^{-1}, thus indicate that the carboxylate ligand function as monodentate ligand under the conditions employed[10]. The bonds at 458 cm^{-1} and 569 cm^{-1} can be assigned to ν (Sn-O) and v(Sn-C) vibration, respectively.

3.2. ^{1}HNMR Spectra

The chemical shifts of the ^{1}H of pyridine ring exhibit signals at region 7.99-8.88 ppm as multiplet, which are considerably shifted towards higher frequencies with respect to that corresponding free carboxylate group[11]. The ^{1}H NMR spectra of the compound show that the chemical shifts of the protons on the phenyl group exhibit signals at 6.78-7.37 ppm as multiplet.

3.3. CrystalStructure

The molecular structure and One-dimensional chain structure are illustrated in Figure 1 and Figure 2, respectively.

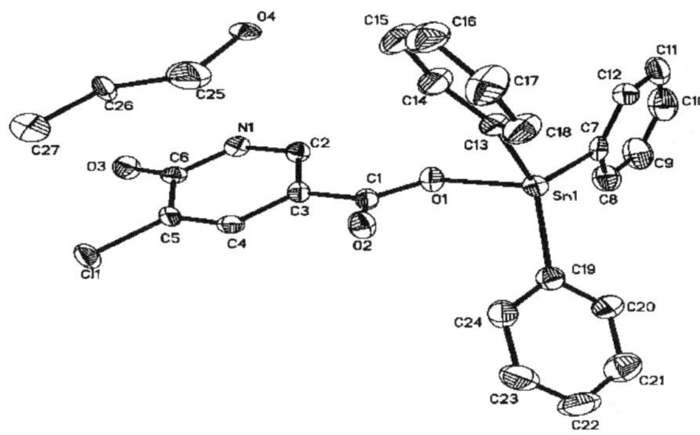

Figure 1. Molecular structural unit of the compound (probability of ellipsoid is 30%, hydrogen atoms have been omitted for clarity).

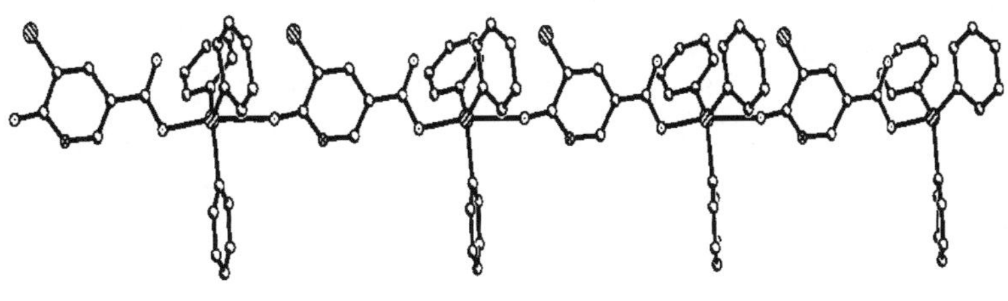

Figure 2. One-dimensional chain structure of the compound.

As can be seen from Figures 1 and 2, the crystal structure of the compound possess unequivocally polymeric structure, but these structures are differ from compounds [$(PhCH_2)_3Sn(O_2CCH_3)$][12] and [$Ph_3Sn(OCOC_5H_4NO)]_4$•$3(CH_2Cl_2)$[13]. Each tin atom is five-coordinated with three phenyl carbon atoms and two oxygen atoms one from carboxylate and other from the phenolic hydroxide.

The intramolecular Sn1-O1 bond length of 2.138(6) Å is longer than $^nC_4H_9SnCl_2(C_{13}H_{11}N_2O_3S)$(2.053 nm)[14], but is shorter than those of $[Ph_3Sn(OCOC_5H_4NO)]_n$ (0.2174 Å)[15]. The Sn1-O3#1 bond length is 2.662(8) Å, longer than the sum of the covalent radii of Sn and O (2.07 Å), but well within the sum of the Van der waals' radii for Sn and O of 2.80 Å, so it should be considered as a bonding interaction. The angles of C13-Sn1-O1 95.7(3)°, C19-Sn1-O1 104.6(3)°, are greater than 90°;but the angles of C19-Sn1-O3#1 83.9(3)°, C7-Sn1-O3#1 81.9(3)° and C13-Sn1-O3#1 83.9(3)° are deviate from 90°, So on the side of the Sn-O bond of the phenyl groups are moved away from carboxylate group. The sum of the C19-Sn1-C7 116.8(3)°, C19-Sn1-C13 122.3(4)° and C7-Sn1-

C13 116.8(4)° bond angles is 355.9°, which shows that the atoms of Sn1, C7, C13 and C19 are not co-planar. The angle of O1-Sn1-O3#1 [169.9(2)°], deviate from the linear angle, for which the coordination at the tin atom is in a distorted trigonal bipyramidal polyhedron.

3.4. Antitumor activity

The antitumor activity was in vitro according to the SRB methods[16]. The compound was tested in vitro against to human hepatocellular carcinoma (SMMC-7721) and muose pmh adenomaly (P388), which displayed quite high activity with inhibition value IC_{50} being 24, 43 ng/mL, respectively. In vitro biological assays suggest that the compound is still rather active against selected tumor cell lines, and the compound exhibit cytotoxic effects against the two cell lines varying in the order SMMC-7721 > P388. The compound displayed the high in vitro antitumor activities, which were more active than clinically widely used cisplatin did.

In summary, a one-dimensional linear polymer has been obtained by the reaction of chlorotriphenyltin with 5-chloro-6-hydroxynicotinic acid. The compound exhibit high antitumor activity. The synthesis of the compound confirms that it is crucial to choose an appropriate carboxylate ligand for the formation of one-dimensional linear polymers. Further studies on antitumor activity of the compound are under way.

References

[1] Wu Y B , Li B W, Li F X,Jian-Wei Xue J W and Lv Z P 2018 *Can. J. Chem* **96** 447-452. DOI:10.1139/cjc-2017-0612
[2] Roner M R, Charles E. Carraher C E, J R , Kimberly Shahi K and Barot G 2011 *Materials***4** 991-1012.DOI:10.3390/ma4060991
[3] Kepert C J, Phillips AE, Goodwin A L, Halder G J and Southon P D 2008 *Angew. Chem. IntEdit* **47** 1396-1399. DOI:10.1002/anie.200704421
[4] Johnston P, Carthey N and Hutchings G J 2015 *J. Am. Chem. Soc* **137** 14548-14557.DOI:10.1021/jacs.5b07752
[5] Salem S A, Ali M A and Mohamed H E 2010 *Molecules* **15** 1784-1797.DOI: 10.3390/molecules15031784
[6] Jain R, Singh R and Kaushik N K 2013 *J. Chem* 1-12.DOI: 10.1155/2013/568195
[7] Zacconi F C, Ocampo R A, Podestá J C and Koll L C 2016 *J. Braz. Chem. Soc* **27** 484-492.DOI: 10.5935/0103-5053.20150275
[8] Adeyemi J O and Onwudiwe D C 2018 *Molecules* **23** 2571-2597. DOI: 10.3390/molecules23102571
[9] Sheldrick G M 1997 SHELXL-97, Program for the Refinement of Crystal Structures, University of Göttingen, Germany. DOI: 10.1107/s2053229614024218
[10] Khatkar P, Asija S and Singh N 2017 *J. Serb. Chem. Soc* **82** 13-23. DOI: 10.2298/JSC160429089K
[11] Yin H D, Wang C H and Xing Q J, 2004 *Polyhedron* **23** 1805-1810. DOI:10.1016/j.poly.2004.04.012
[12] Alcock N W and Timms R E. 1968 *J. Chem. Soc. A* 1876-1880.
[13] Gao Z J, Tian G R, Zhao G W and Si C D 2010 *Inorg. Chem. Commun* **13** 461-463.DOI:10.1016/j.inoche.2009.12.026
[14] Yin H D and Chen S W 2006 *J. Organomet. Chem* **691** 3103-3108.DOI: 10.1016/j.jorganchem.2006.03.003
[15] Gao Z J, MU X, Wang G H. 2008 *Chin. J. Inorg. Chem* **24** 1916-1918.
[16] Skehan P, Storeng R, Scudiero D, Monks A,Mcmahon J,Vistica D,Warren J T,Bokesch H, Kenney S and Boyd M R 1990 *J. Natl. Cancer. Inst* **13** 1107-1112.DOI:10.1093/jnci/82.13.1107

Experimental and numerical studies of perforated CFRP laminates under quasi-static tensile load

Dejun Yin[1,2], Jian Zheng[1,3], Chao Xiong[1] and Junhui Yin[1]

[1]Department of Artillery Engineering, Shijiazhuang Campus, Army Engineering University, Shijiazhuang City 050003, China;
[2]Unit 63961 of PLA, Beijing 100012,China

[3]Email: zj3373775258@163.com

Abstract. Mechanical properties and failure modes of perforated CFRP laminate specimens under quasi-static tensile load were studied by means of experimental test and numerical simulation. Tensile tests of perforated CFRP laminates were conducted on an INSTRON electronic universal testing machine, and the progressive damage model based on 2D Hashin failure criterion was established by using the ABAQUS/Explicit software. The results showed that the discrepancy of ultimate tensile load basis on experimental test and numerical simulation was acceptable, which was less than 6.0 %. The numerical simulation method and results provide a theoretical reference for the design and application of perforated CFRP laminates.

1. Introduction

As a typical representative of advanced composite materials, carbon fiber reinforced plastic (CFRP) has been widely used in aerospace, automobile, high-speed train and other fields attributed to its characteristics of low density, high specific modulus, high specific strength and liable to design compared with traditional materials [1-6]. Specifically, the application of CFRP in Boeing 787 reaches 50%, and that in the wing and fuselage of F35 fighter takes up as much as 30%. Due to the needs of inspection, disassembly and maintenance, CFRP structures inevitably need some holes in the process of use, such as the side wall panel of the aircraft needs holes for installing doors and windows, and the internal bulkhead needs holes to make the pipeline pass through.

Early work in the area, a large number of researchers have carried out a lot of research on the strength and failure of perforated composite structures[7]. For example, Lekhnitskii [8] proposed a theoretical formula for calculating the stress concentration at the hole edge of orthotropic plates. Nuismer [9] proposed the point stress criterion and the average stress criterion for predicting the tensile strength of laminated plates with holes. Chang [10] proposed a progressive damage model for laminates with notches under tensile loading, which was used to predict the failure under tensile or compressive loading. Su et al. [11] developed a progressive damage model based on ABAQUS UMAT, and studied the effect of opening size on CFRP laminates with openings under compression load. Xu et al. [12] found that the cohesive zone model could effectively predict the mechanical properties of CFRP specimens with holes, and the error was within 10%. Wang et al. [13] found that the deformation capacity, axial stiffness and critical load of perforated glass fiber reinforced plastic (GFRP) pipes under compression load were significantly affected by different parameters.

In the practical application of CFRP structures, it is often necessary to open holes to meet various functional requirements. The purpose of this study is to understand the influence of perforated structure on the bearing capacity and failure characteristics of the CFRP laminates. In this paper, the mechanical properties and failure modes of perforated CFRP laminate are studied by experimental test and numerical simulation, which provides an important theoretical reference for the engineering application of CFRP structures.

2. Experimental methods

2.1. Sample preparation

Domestic T300 carbon fiber epoxy unidirectional prepreg was chosen to fabricate the CFRP laminates by the bladder molding process and the detailed curing process curve is shown in Figure 1. The thickness of CFRP laminates is 2.4 mm (16 plies) with a stacking sequence of $[0/90/45/-45]_{2s}$. The CFRP laminates were processed into perforated tensile specimens using CNC engraving machine based on ASTM Standard D7137-07, and the schematic diagram of geometric dimension for CFRP laminate specimen is shown in Figure 2.

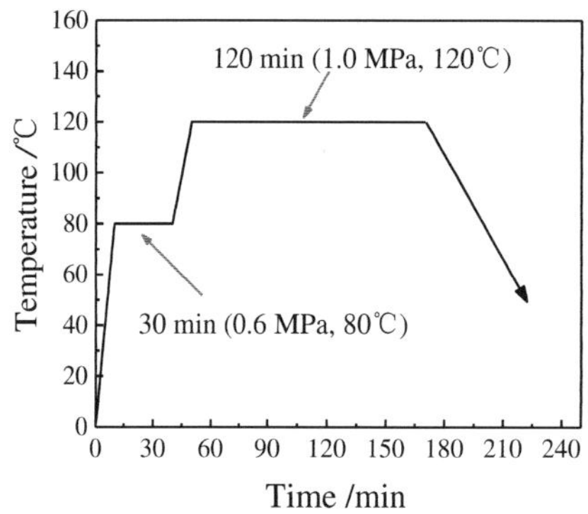

Figure 1. Curing process curve of CFRP laminates.

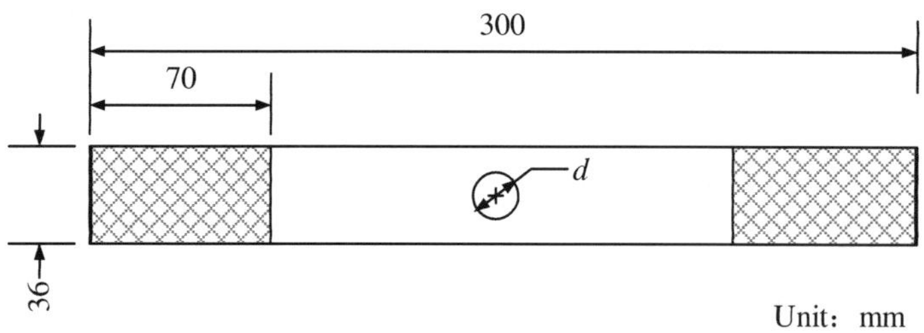

Figure 2. Schematic diagram of specimen geometric dimension.

2.2. Testing procedure

The quasi-static tensile tests were carried out on the electronic universal testing machine (INSTRON-5982, U.S.) with measuring range of 100 kN at room temperature, as shown in Figure 3. The displacement loading mode was performed, and the loading rate is 2 mm/min. The load data was recorded automatically by the testing machine, and the displacement data was measured by the 3D digital image correlation (DIC) test system (PMLAB, China). Detailed test and analysis methods can refer to the existing literatures [4, 14]. To ensure the reliability of the data, five samples were prepared and tested.

Figure 3. Schematic diagram of the experimental apparatus.

3. Numerical modelling

The load-displacement curves of perforated CFRP laminate specimens under tensile load can be obtained through the experimental study. However, it is difficult to observe internal damage of CFRP laminate specimens due to the limitation of characterization techniques. In order to study the internal damage mechanism and failure mode of perforated CFRP laminate specimens, the finite element simulation analysis was carried out based on ABAQUS/explicit solver.

3.1. Failure modes and failure criteria

The failure modes of CFRP laminate were strongly dependent on the geometry, loading direction and ply direction. The typical failure modes can be divided into in-plane failure modes and transverse failure modes. The anisotropic failure modes in ABAQUS were based on the research work of Hashin et al. [15, 16]. The damage initiation criterion of CFRP laminate was based on the Hashin failure criterion. For single-layer CFRP, four different failure modes are considered: fiber tensile fracture, fiber compression buckling and torsion, matrix cracking under transverse tension and shear force, matrix fracture under transverse pressure and shear force. According to the theory, damage begins when one of the four values of fiber tension (F_{ft}), fiber compression (F_{fc}), matrix tension (F_{mt}) or matrix compression (F_{mc}) reaches the breaking strength of CFRP.

Fiber tension ($\sigma_{11} \geq 0$):

$$F_{\mathrm{ft}} = \left(\frac{\sigma_{11}}{X_{\mathrm{T}}}\right)^2 + \alpha\left(\frac{\tau_{12}}{S_{\mathrm{L}}}\right)^2 \tag{1}$$

Fiber compression ($\sigma_{11} < 0$):

$$F_{\mathrm{fc}} = \left(\frac{\sigma_{11}}{X_{\mathrm{C}}}\right)^2 \tag{2}$$

Matrix tension ($\sigma_{22} \geq 0$):

$$F_{\mathrm{mt}} = \left(\frac{\sigma_{22}}{X_{\mathrm{T}}}\right)^2 + \alpha\left(\frac{\tau_{12}}{S_{\mathrm{L}}}\right)^2 \tag{3}$$

Matrix compression ($\sigma_{22} < 0$):

$$F_{\mathrm{mc}} = \left(\frac{\sigma_{22}}{2S_{\mathrm{T}}}\right)^2 + \left[\left(\frac{Y_C}{2S_{\mathrm{T}}}\right)^2 - 1\right]\frac{\sigma_{22}}{Y_{\mathrm{C}}} + \left(\frac{\tau_{12}}{S_{\mathrm{L}}}\right)^2 \tag{4}$$

where σ_{11}, σ_{22}, and τ_{12} denote the components of stress tensor, while α is the contribution coefficient that determines the role of shear stress in fiber tensile failure. X_T and X_C are the tensile strength and compressive strength of the fiber, respectively. Y_T and Y_C represent the tensile strength and compressive strength of the matrix, respectively. S_T and S_C denote the longitudinal and transverse shear strength, respectively.

σ_{ij} is applied to evaluate the initiation criteria, and the calculation formula is as follows:

$$\sigma_{ij} = \boldsymbol{M}\sigma \tag{5}$$

where σ_{ij} is the nominal stress, and $\boldsymbol{M}$ is the damage operator [17].

$$M = \begin{bmatrix} \frac{1}{(1-d_f)} & 0 & 0 \\ 0 & \frac{1}{(1-d_m)} & 0 \\ 0 & 0 & \frac{1}{(1-d_s)} \end{bmatrix} \tag{6}$$

d_f, d_m, and d_s are the internal variables of fiber failure, matrix failure, and shear failure, respectively. d_{ft}, d_{fc}, d_{mt}, and d_{mc} are the rupture damage variables of fiber tension, fiber compression, matrix tension, and matrix compression failure previously discussed, respectively. Details are given below.

$$d_f = \begin{cases} d_{ft} & \sigma_{11} \geq 0 \\ d_{fc} & \sigma_{11} < 0 \end{cases} \tag{7}$$

$$d_f = \begin{cases} d_{mt} & \sigma_{22} \geq 0 \\ d_{mc} & \sigma_{22} < 0 \end{cases} \tag{8}$$

$$d_s = 1-(1-d_{ft})(1-d_{fc})(1-d_{mt})(1-d_{mc}) \tag{9}$$

Detailed mechanical parameters of CFRP laminates are shown in Tables 1 and 2.

Table 1. Elastic modulus and Poisson's ratio of CFRP.

Density	Elastic modulus (GPa)		Poisson's ratio		
(kg/ m^3)	E_{11}	E_{22}	ν_{12}	ν_{13}	ν_{23}
1614	135	8.8	0.33	0.33	0.33

Table 2. Strength and damage evolution of CFRP.

Strength (MPa)						Damage evolution (mJ/mm^2)			
X_T	X_C	Y_T	Y_C	S_T	S_L	G_{ft}	G_{fc}	G_{mt}	G_{mc}
1548	1226	55.5	232	89.9	89.9	50.5	30.5	0.22	1.1

3.2. Finite element analysis model

Full-scale finite element model of perforated CFRP laminate specimen under tensile load was established by ABAQUS. The finite element model was consistent with the specimen size and the boundary condition was consistent with the experimental clamping conditions. The left end of the model was constrained by fixed end, and the right end was loaded with displacement. The SC8R element was used to mesh the CFRP laminates, and the stress concentration area of the circular hole was refined, as shown in Figure 4.

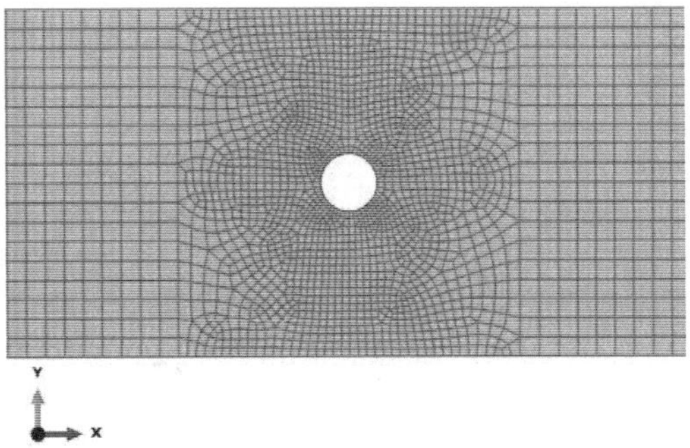

Figure 4. Refinement mesh near the circular hole of CFRP laminate specimen.

4. Results and discussion

4.1. Load-displacement histories

Figure 5 plots the internal energy-time curve and kinetic energy-time curve in the numerical simulation of perforated CFRP laminate tensile specimen. It can be seen that the internal energy of the whole model is far greater than the kinetic energy before the failure of the specimen (t<0.35 ms). Therefore, the dynamic effect of using ABAQUS/explicit solver to simulate quasi-static tensile test can be ignored.

In order to verify the correctness of the progressive damage model, the simulation results were compared with the experimental results. Figure 6 shows the comparison of load-displacement curves obtained by experimental measurement and numerical prediction for CFRP laminates with 6 mm circular hole under tensile load. The displacement in the experiment was the displacement of the fixture measured by 3D-DIC test system. It can be seen that the experimental value of the tensile load was 31.82 kN, and the simulation value was 33.46 kN which was slightly higher than the experimental value. It is interesting to note that the failure displacement measured by 3D-DIC test system was slightly higher than that in the simulation. The relative errors were less than 6 %, which was within the allowable range. Generally speaking, the numerical prediction was in good agreement with the experimental measurement, which showed the correctness of the numerical model.

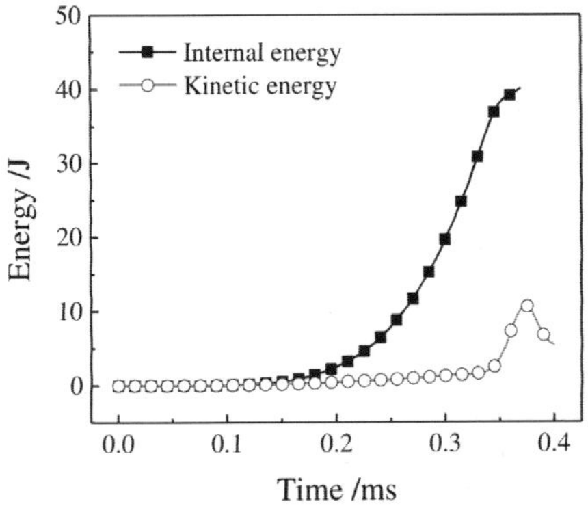

Figure 5. Comparison of energy-time curves in the numerical simulation of tensile test.

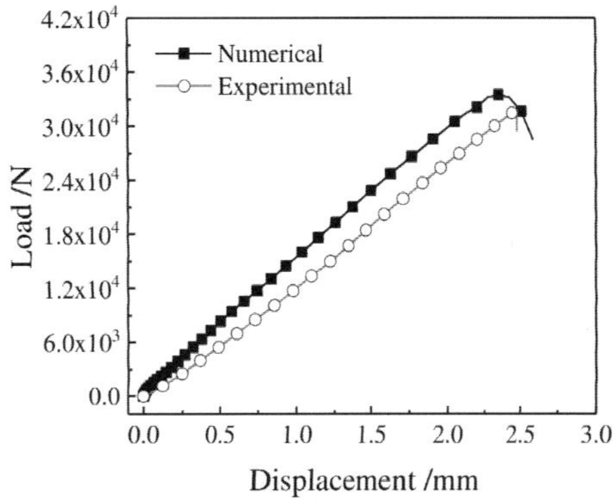

Figure 6. Comparison of load-displacement curves in tensile test.

IOP Conf. Series: Materials Science and Engineering 1040 (2021) 012011 doi:10.1088/1757-899X/1040/1/012011

4.2. Failure modes

The existence of holes results in local stress concentration of the structure, which leads to a significant reduction in the bearing capacity of the structure, and directly affects the service life and safety of the structure. The initial damage and evolution of CFRP are more complex because of the damage of stress concentration and fiber continuity at the opening. Therefore, it is very important for the application of CFRP structures to accurately predict the effect of opening on the failure modes for CFRP specimen. When the load reached ultimate tensile load 33.46 kN, the specimen eventually failed. Four failure modes of fiber tension, fiber compression, matrix tension and matrix compression obtained by progressive damage model are shown in Figure 7. It is interesting to note that the matrix crack at the edge of the hole accelerated the fiber fracture. This is due to the rapid expansion of 90° layer and ±45° layer matrix crack, which made the matrix lose the bearing capacity, and released the stress concentration at the hole edge.

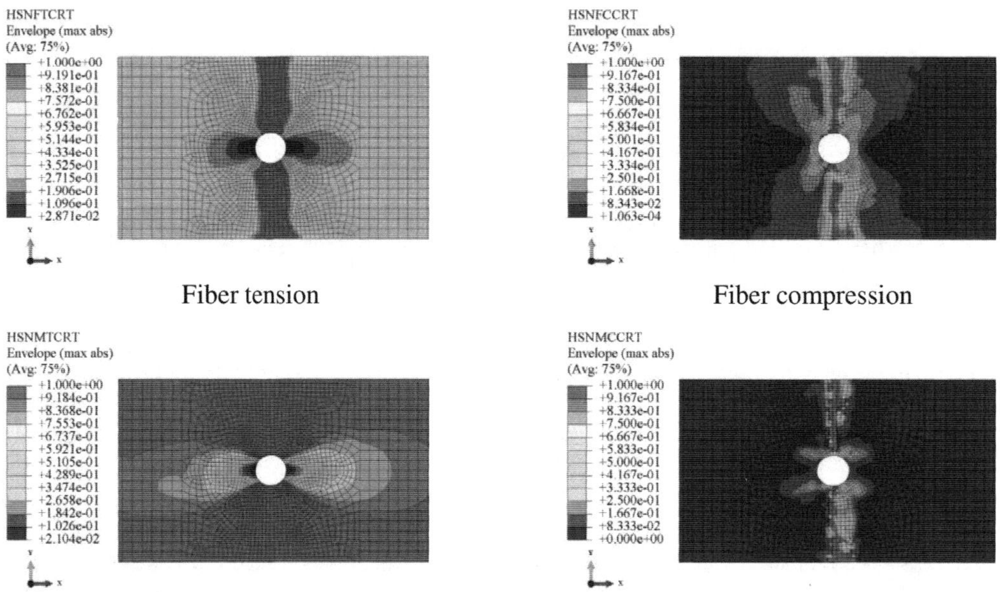

Figure 7. Failure modes of perforated CFRP laminate specimen at ultimate tensile load.

5. Conclusions

The load-displacement curves of perforated CFRP laminate specimens under tensile load were obtained by combining the INSTRON universal testing machine and 3D-DIC testing system. The progressive damage finite element model of perforated CFRP laminate specimen was established by using ABAQUS/Explicit based on 2D Hashin criterion. The tensile behavior of perforated CFRP laminate specimens was analyzed and the residual strength was predicted by the finite element mode. The maximum error between the experimental results and the predictions of tensile strength was less than 6.0%, which verified the effectiveness of the simulation calculation method. The initial damage and evolution process of perforated CFRP laminate specimens were analyzed through the finite element simulation. The finite element model established in this paper can accurately predict the behavior of damage initiation, expansion and final failure of CFRP laminate.

References

[1] Zhang Y, Zhang L L. 2019 Application of carbon fiber composites in modern automotive industry [J]. *Fiber Composites* **53(3)** 53-57

[2] Liu Q, Lin Y, Zong Z, *et al.* 2013 Lightweight design of carbon twill weave fabric composite

body structure for electric vehicle [J]. *Composite Structures* **97** 231-238
[3] Luo X M, Luo Y F. 2011 The application of several high performance fibers and composite materials in high speed train both in abroad and home [J]. *Hi-Tech Fiber & Application* **36(5)** 33-41
[4] Yin D J, Zheng J, Xiong C, *et al.* 2020 Analytical, numerical, and experimental investigations on transverse bending responses of CFRP square tube filled with aluminum foam [J]. *Mathematical Problems in Engineering* **2020 Article ID 8764349** 17
[5] Yin D J, Zheng J, Xiong C, *et al.* 2019 Compressive behaviour and failure mechanisms of GFRP composite at high strain rates [J]. *Journal of Beijing Institute of Technology (English Edition)* **28(1)** 184-190
[6] Huang X G, Yin C, Huang J, *et al.* 2016 Hypervelocity impact of TiB_2-based composites as front bumpers for space shield applications [J]. *Materials and Design* **97** 473-482.
[7] Stamopoulos A G, Tserpes K I, Pantelakis S G. 2018 Multiscale finite element prediction of shear and flexural properties of porous CFRP laminates utilizing X-ray CT data [J]. *Theoretical and Applied Fracture Mechanics* **97** 303-313
[8] Lekhnitskii S G. 1981 Theory of elasticity of an anisotropic body *[M]. San Francisco: Mir Publisher*
[9] Nuismer R J, Whitney J M. 1975 Uniaxial failure of composite laminates containing stress concentrations [J]. *Fracture Mechanics of Composites* **49(3)** 117-142
[10] Chang K Y, Liu S. 1991 Damage tolerance of laminated composites containing an open hole and subjected to tensile loading [J]. *Journal of Composite Materials* **25(3)** 274-301
[11] Su Z C, Tay T E, Ridha M, et al. 2015 Progressive damage modeling of open-hole composite laminates under compression [J]. *Composite Structures* **122** 507-517
[12] Xu W, Thorsson S I, Waas A M. 2015 Experimental and numerical study on cross-ply woven textile composite with notches and cracks[J]. *Composite Structures* **132** 816-824
[13] Wang W, Sheikh M N, Hadi M N S. 2015 Behaviour of perforated GFRP tubes under axial compression [J]. *Thin-Walled Structures* **95** 88-100
[14] Hu J S, Zhang K F, Yang Q D, *et al.* 2018 An experimental study on mechanical response of single-lap bolted CFRP composite interference-fit joints[J]. *Composites Structures* **196** 76-88
[15] Hashin Z, Rotem A. 1973 A fatigue failure criterion for fiber reinforced materials [J]. *Journal of Composite Materials* **4** 448-464
[16] Hashin Z. 1980 Failure criteria for unidirectional fiber composites [J]. *Journal of Applied Mechanics* **47(2)** 329-334.
[17] Lee S M , Seo H Y, Kang C G, *et al.* 2016 Properties of CR340/CFRP composites through three point bending test by using theoretical and experimental methods [J]. *International Journal of Precision Engineering and Manufacturing-Green Technology* **3(4)** 359-365

Thermal deformation behavior of Stellite 6B alloy

Zhang Yawei[1,2], Zhang Shixiao[2] and Lu Xudong[1,2,3]

[1]Beijing Key Laboratory of Advanced High Temperature Materials, Central Iron and Steel Research Institute, Beijing 100081, China;
[2]Beijing Gaona Aero Material CO., LTD., Beijing 100081, China

[3]Email: lxdong0700@hotmail.com

Abstract. Stellite 6B is carbide-hardened Co-based superalloy. It finds extensive application in the harsh industrial environment, because of excellent wear characteristics, high hardness, good corrosion resistance, and superior mechanical properties. Hot compression tests were performed on Stellite 6B alloy to study high temperature dynamic recrystallization and microstructure evolution during thermal deformation. The tests were performed in the temperature 1000 °C, 1050 °C, 1100 °C, 1150 °C and 1200 °C and at the strain rates of 0.01 s^{-1}, 0.1 s^{-1}, 1 s^{-1} and 10 s^{-1}. Stress-strain curves, and the microstructural evolution of the Stellite 6B alloy were investigated. The results showed that the recrystallization was nucleated at the carbides, original grain boundaries and twin grain boundaries, the dynamic recovery process is inhibited, and the softening effect by dynamic recrystallization is more significant.

1. Introduction

Stellite 6B is a typical carbide-hardened Co-based wear-resistant superalloy, and it has been widely used in cement and steel equipment, rock crushing rollers, transmission devices, and other industrial fields, especially in sleeves that cannot be fully lubricated [1], because of their corrosion resistance, high-strength, and ability to retain hardness at high temperatures [2]. The typical microstructure of this kind of alloy is many primary carbides dispersed in the cobalt-rich solid solution matrix. Dynamic recrystallization is the key factor to determine the microstructure and mechanical behavior of nickel-based superalloys during thermal deformation. Many scholars have studied the thermal deformation constitutive relationship, microstructure evolution and nucleation mode of dynamic recrystallization of various nickel-based superalloys and discussed the effects of precipitated phase and annealing twins. Poliak et al. established a model of dynamic recrystallization [3]. Gourdet et al. proposed a simple continuous dynamic recrystallization (CDRX) model [4]. Lee et al. investigated the dynamic deformation behavior, fracture characteristics and microstructural evolution of Inconel 690 alloy under high speed loading conditions [5]. Poelt et al. studied the dynamic recrystallization of the Ni-base alloy in a transient state by light microscopy and electron backscatter diffraction (EBSD) [6]. Cai et al. conducted hot compression testing of a Ni-based superalloy using processing maps [7]. Liu, J. et al. proposed a new kinetics model of dynamic recrystallization (DRX), which divided the development process of DRX into three phases: slow-beginning phase, rapid-increasing phase and slow-rising-to-balance phase [8]. Amiri et al. studied the hot deformation behavior of the Waspaloy, the processing map was developed based on experimental data, showing stable domain and one instable domain [9]. Azarbarmas et al. investigated the microstructural evolution of IN718 by hot compression tests. The results indicated that continuous dynamic recrystallization (CDRX) is promoted at higher strains, large strain rates and lower temperatures, both DDRX and CDRX can occur simultaneously during the hot deformation of IN718 under certain conditions [10].

Comparing to much harder cast Co-based alloy, the wrought structure has excellent abrasion wear resistance. For the wrought Co-based alloy, dynamic recrystallization (DRX) is considered as the most important softening mechanism in thermal processing, which is beneficial to obtain uniform fine microstructure and improve mechanical properties of products. However, there are still only a few reports in the literature so far concerning the dynamic recrystallization behavior and microstructure evolution of Stellite 6B alloy during thermal deformation. In this paper, hot compression tests were performed on Stellite 6B alloy to study high temperature dynamic recrystallization behavior and microstructure evolution during thermal deformation at 1000 °C、 1100 °C and 1200 °C.

2. Experimental procedures

The chemical compositions (wt.%) of Stellite 6B used in this investigation are as follows: C 1.2; Cr 30; Mo 1.2; W 4.5; Ni 2.0; Si 1.0; Mn 1.0; Fe<3.0; Co balance. Several 12.5mm thick sheets were cut from an as-received wrought bar (40 mm in diameter) in the direction of cross section. Cylindrical specimens, with a diameter of 8 mm and a height of 12 mm, were machined from the center part to ensure the uniformity of microstructure for all the specimens. The specimens were then solution treated at 1220 °C for 30 min before compression tests.

The isothermal compression tests were performed on the thermomechanical simulator Gleeble-3800 at the temperatures ranging from 1000 °C to 1200 °C at the intervals of 50 °C and at the strain rates of 0.01 s^{-1}, 0.1 s^{-1}, 1 s^{-1} and 10 s^{-1}. Sections were cut from the specimens along the longitudinal axis, and then the center areas of the specimens were used for metallographic analysis. The samples for EBSD evaluation were ground with SiC papers, mechanically polished with diamond paste, and then electrolytically polished with 20 pct solution of H_2SO_4 in methanol.

A JSM-7200F type electron backscatter diffraction (EBSD) detector was used to analyze the microstructural evolution of the compression specimens. All postprocessing of maps, the grain size analysis and misorientation angle calculation, were performed in HKL Channel5 software. Noise reduction was performed on EBSD patterns, with removal of wild spikes and zero solutions. The foils for TEM examination were prepared by grinding the specimens to a thickness between 50 and 60μm. Afterwards, the foils were thinned by twin-jet electropolishing using a solution of 10% perchloric acid in ethanol. The TEM examination was conducted on a Tecnai G^2 F20 microscope.

3. Results and discussion

3.1. *True stress-true stain curves*

Figure 1 shows the true stress-true strain curves of Stellite 6B alloy obtained under different temperatures and strain rates. The stress-strain curve appears a single peak near the critical strain, and then the strain softening stage appears. This is a typical characteristic of the flow stress curve of the low stacking fault energy alloy, indicating that DRX phenomenon occurs in the thermal deformation process [3]. In the initial stage of the forming process, the flow stress abruptly increases due to the dominance of work hardening. When deformation exceeds the peak strain, the flow stress decreases with increasing stain as softening caused by DRX. Due to the balance between dynamic softening and work hardening, the flow stress presents a steady-state region.

The flow stresses are dependent on the deformation variables, such as temperature, strain rate and strain. As temperature decreases at the constant strain rate, the overall level of the flow curve enhances correspondingly due to the decrease in the dislocation slip and the diffusion ability of grain boundaries for the annihilation of dislocations. At the same deformation temperature, the flow stress increases with the increase of strain rate.This is mainly because the lower strain rate provides longer dislocation slip and dislocation annihilation time, thus reducing the flow stress.

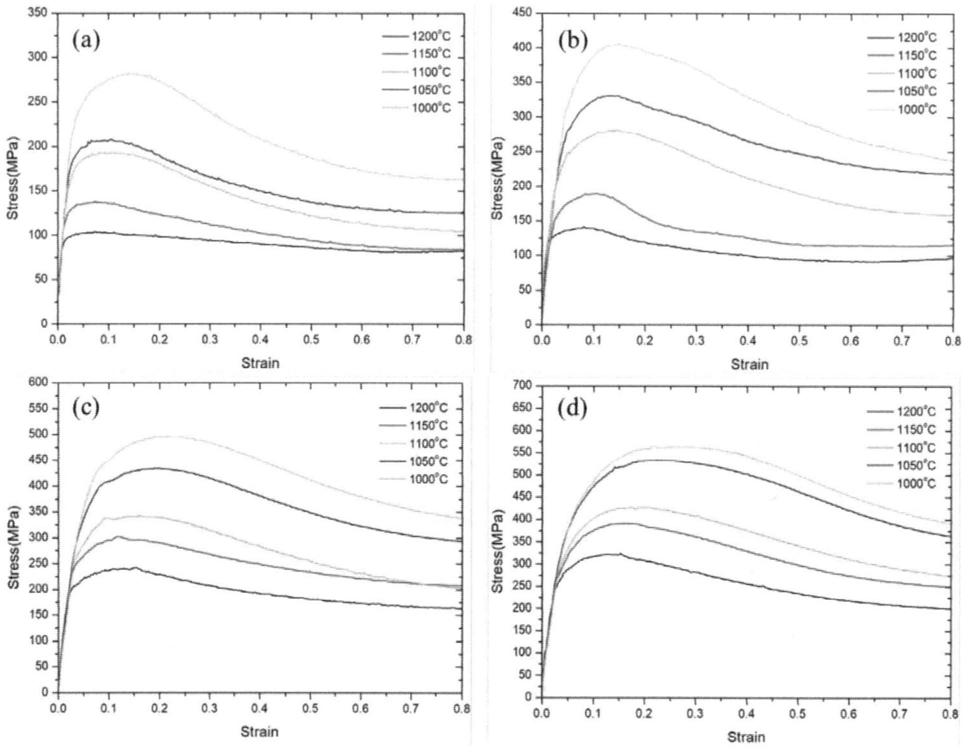

Figure 1. Flow stress curves of Stellite 6B alloy at different temperatures and strain rates of (a) 0.01s^{-1}, (b) 0.1 s^{-1}, (c) 1 s^{-1}, (d) 10 s^{-1}.

Peak stress is an important indicator in the process of thermal processing, which determines the setting of the power parameters and the microstructure evolution. The peak stress of Stellite 6B alloy in different deformation temperature and strain rate is shown in Figure 2, and it can be observed that the peak stress decreases with the decreasing strain rates and the increasing temperature, which is a conventional behavior and consistent with other reports for superalloys [11]. It is worth mentioning that the peak strain (the one associated to the peak stress), which is usually associated to the onset of DRX, is also decreases with decreasing strain rates and the increasing temperature.

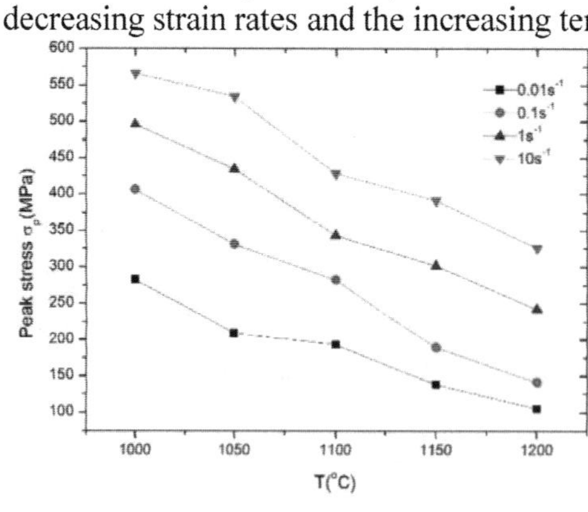

Figure 2. Schematic of the dependence of peak stress σ_p on strain rates and temperatures.

3.2. Dynamic model of recrystallization

Generally, JMAK model is used to describe the DRX kinetics dynamic recrystallization of alloys, that is, the DRX volume fraction is expressed as a function of deformation time as follows:

$$X_{DRX} = 1 - \exp\left[-0.693\left(\frac{t}{t_{0.5}}\right)^{k}\right]$$

where X_{DRX} is the dynamically recrystallized volume fraction, $t_{0.5}$ is the time for 50% DRX, t is the time for X_{DRX} to occur. A lot of works needs to do to calculate the recrystallization fraction of the alloy after rapid cooling under different deformation conditions at multiple time intervals. Finally, the as-obtained DRX kinetic model of the Stellite 6B alloy is described as:

$$\begin{cases} X_{DRX} = 1 - exp\left[-\left(\frac{\varepsilon - \varepsilon_c}{\varepsilon^*}\right)^{m}\right] \\ m = 1.12\left(\frac{Z}{A}\right)^{-0.08} \\ \varepsilon_c = 0.001477Z^{0.10001} \\ \varepsilon^* = 0.003305Z^{0.10834} \end{cases}$$

3.3. The microstructural evolution

The evolution of DRX grain at high temperature, such as nucleation and growth, is usually studied by characterizing grain structure and its internal substructure. The formation of recrystallization grains in the deformation process is completed through the two basic processes of nucleation and growth, so the microstructural of dynamic recrystallization mainly depends on these two processes. This section mainly studies the microstructure evolution of Stellite 6B alloy under the condition of hot compression deformation. EBSD was used to obtain detailed data on orientation error, grain boundary characteristics, grain size, recrystallization mechanism and recrystallization fraction.

Figure 3 shows the microstructure at different temperatures and strains of Stellite 6B alloy, the constant strain rate of $1s^{-1}$. The dynamic recrystallization of the alloy occurred at all conditions of thermal compression deformation, and a large number of twin boundaries can be observed. With the increasing of strain at all temperatures, the number of recrystallization nucleation increases rapidly. As can be seen from the Figure 3(a), when the deformation temperature is 1000 ℃ and the deformation amount is 10%, dynamic recrystallization begins at carbides. When the deformation amount increases to 20%, the recrystallization near the carbides is very sufficient, and a large number of bulging nuclei occurs at the corrugations or serrations grain boundary, as shown in Figure 3(b). When the deformation amount increases to 40%, the faction of dynamic recrystallization is plenty as shown in Figure 3(c) which is corresponds true strain of 0.51, exceeding the maximum softening strain of 0.42. However, the growth rate of the grain is slow at the lower temperature, the grain is obviously refined, and the average grain size is about 3-5 μm. When the deformation temperature is 1100℃ and the deformation amount is 10%, the strain exceeds the critical strain ε_c, the recrystallization near the carbides is remarkable. There was no significant change in the microstructure from 20% to 40%. Figure 3(g), (h), (i) shows that at 1200℃, a large amount of dynamic recrystallization can occur with a small deformation, and the recrystallization grains grow faster after nucleation. When the deformation amount is 20%, dynamic recrystallization basically completed, when the deformation amount increases to 40%, recrystallized grains grow obviously. As the strain increases, the two opposite processes overlap: one is the work hardening of the recrystallized grains; the other one is that the recrystallized grains continue to grow, and the average grain size is about 20-30 μm. Figure 3(c), (f), (i) shows that the dynamic steady-state grain size increases significantly with the increase of temperature.

IOP Conf. Series: Materials Science and Engineering 1040 (2021) 012012 doi:10.1088/1757-899X/1040/1/012012

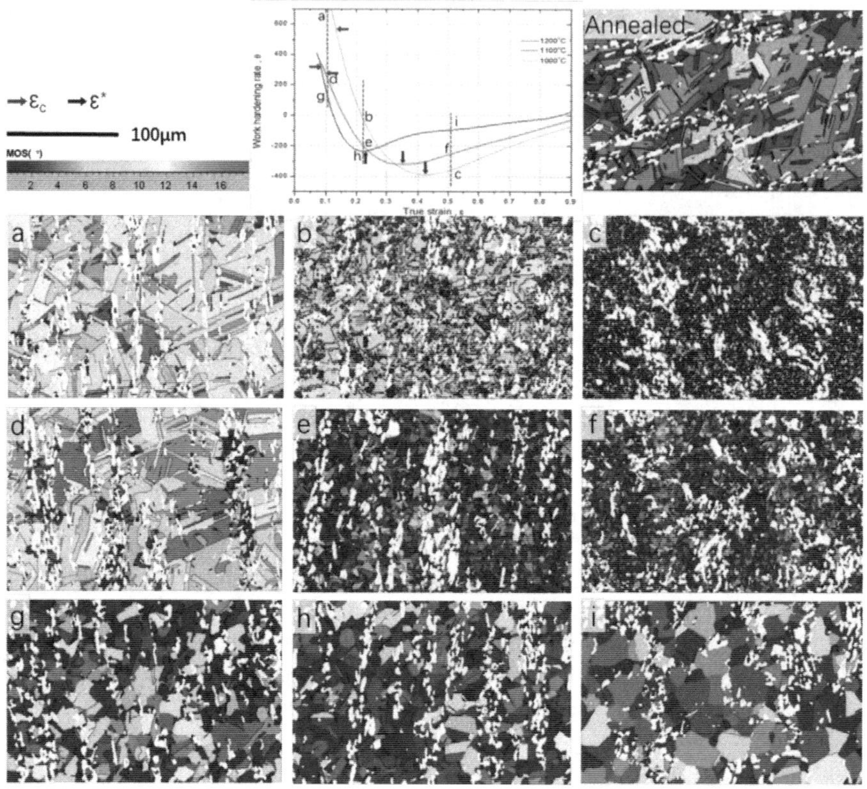

Figure 3. Maximum orientation spread (MOS) maps at various deformation conditions (a) 1000℃, 10%; (b) 1000℃, 20%; (c) 1000℃, 40%; (d) 1100℃, 10%; (e) 1100℃, 20%; (f) 1100℃, 40%; (g) 1200℃, 10%; (h) 1200℃, 20%; (i) 1200℃, 40%.

Schematic illustration of the microstructural evolution at various deformation conditions was summarized as shown in Figure 4. It can also be seen from Figure 3 and Figure 4, with the increase of deformation, many deformation twins are generated in the recrystallized grain during the deformation at 1000 ℃, and the grain structure is significantly refined. At 1100 ℃, with the continuation of the recrystallization process, some twins within the regenerated grain disappeared with the migration of the large Angle grain boundary. At 1200 ℃, the recrystallized grain grew significantly, and the number of twins decreased significantly. With the increase of deformation, recrystallization nucleates at the bending of the original grain boundaries and twin boundaries. Twins play an important role in the dynamic recrystallization of Stellite 6B alloy. The recrystallization grain size of the alloy is determined by the disappearance and proliferation of twins.

As illustrated in Figure 5, in the annealed sample, many of them have Σ3 twin relationship. Some of these twin boundaries are coherent (straight). As shown in Figure 6, when the deformation temperature is 1000℃, the dynamic recrystallization nucleation has not started, and the small-angle grain boundary proliferates in large quantities under small deformation. When the deformation reaches 40%, the recrystallization occurs in the form of twin, and most of the new generation of borders have Σ3 twin relationship. Conversely, when deformation temperature is 1200 ℃, the new DRX grains start to consume the deformed microstructure and the fraction of high angle ones increasing. Different from the above two cases, when the deformation temperature is 1100 ℃, with the increase of the deformation, the large Angle grain boundary gradually increases, and the twin boundary first increases and then decreases.

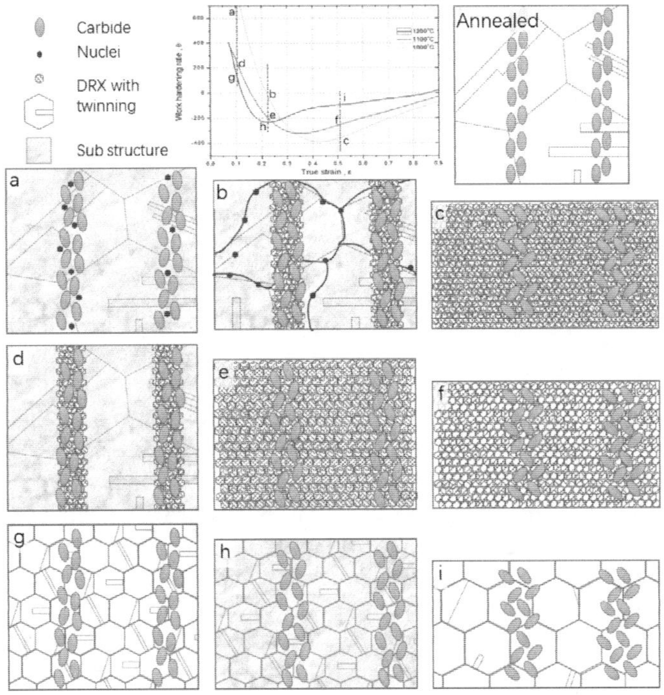

Figure 4. Schematic illustration of the microstructural evolution at various deformation conditions (a) 1000℃, 10%; (b) 1000℃, 20%; (c) 1000℃, 40%; (d) 1100℃, 10%; (e) 1100℃, 20%; (f) 1100℃, 40%; (g) 1200℃, 10%; (h) 1200℃, 20%; (i) 1200℃, 40%.

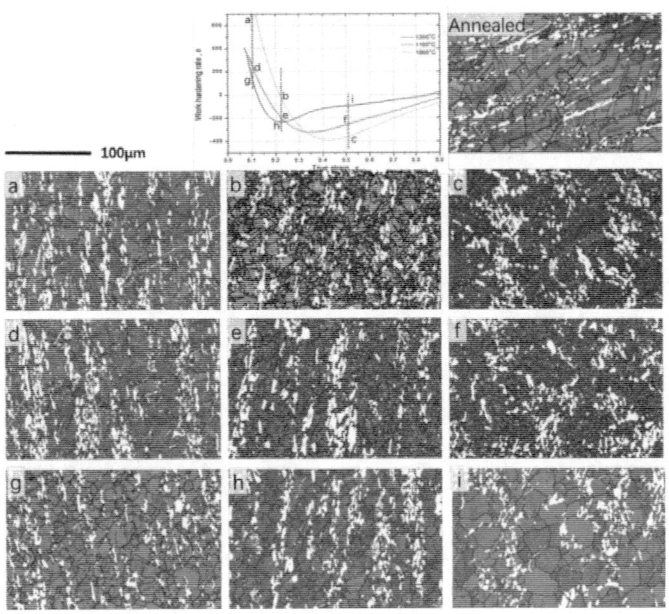

Figure 5. OIM band contrast map at various deformation conditions (a) 1000℃, 10%; (b) 1000℃, 20%; (c) 1000℃, 40%; (d) 1100℃, 10%; (e) 1100℃, 20%; (f) 1100℃, 40%; (g) 1200℃, 10%; (h) 1200℃, 20%; (i) 1200℃, 40%.

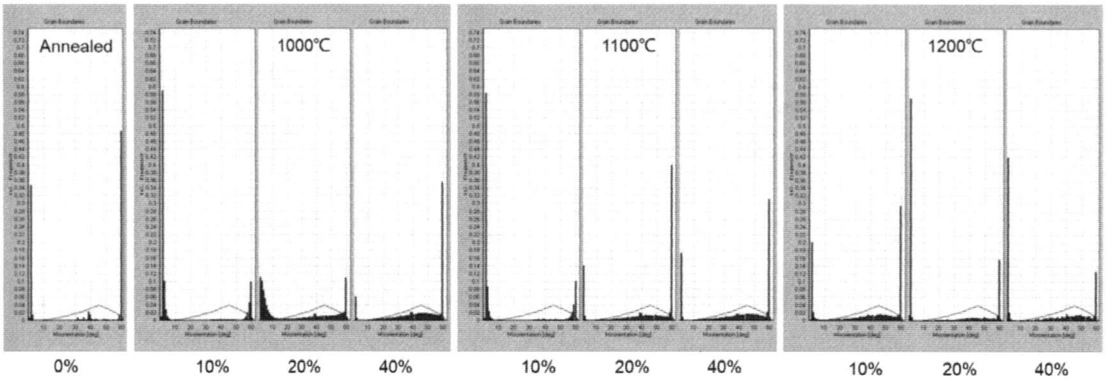

Figure 6. Boundary misorientation angle distribution at various deformation conditions.

In addition, it is conducive to the formation of recrystallize core that the dislocations piled up on the interface of the hard-deformable carbides with a volume fraction of more than 20%. Under the condition of thermal deformation, the multi-series slip in the matrix of Stellite 6B alloy is activated, and a large number of dislocations slip into the vicinity of the series-shaped carbides and are pinned, resulting in serious plugging. As the amount of deformation continues to increase, a large number of new subgrain boundaries are generated, which are distributed both near the original grain boundary and inside the grain, because the activation of slip systems in different directions within the grain decompositions the original grain into different parts. As shown in Figure 7(a), the subgrains are merging through that the dislocation on the adjacent subgrain boundaries moves to the surrounding grain boundaries or subgrain boundaries by climbing and sliding mechanisms. With the increasing of deformation, subgrain boundaries disappeared and recrystallized grains formed.

Wang et al. reported that the formation of twins plays an important role in the nucleation and growth of DRX in materials with low stacking fault energy [12]. The grain boundary migrations are frequently accompanied by twining. Grain boundary bulging is the main nucleation method under high temperature or low strain rate deformation conditions. The convex part of curved grain boundary can induce nucleation of new recrystallized grain. The effects of twins on bulging part separation and DRX nucleation are shown in Figure 7(b) and Figure 8. There is evidence that new DRX grains are formed by repeated nucleation and growth of deformation twins at the migration interface during formation and under steady state conditions. Some evidence of this effect can be seen in Figure 7(b).

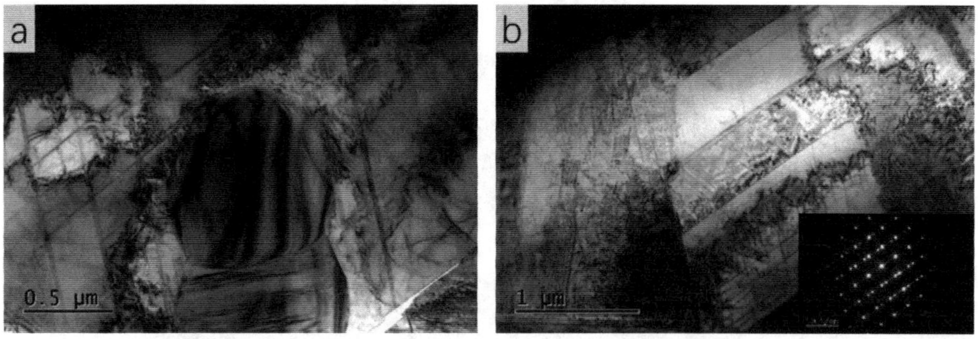

Figure 7. TEM photographs show the recrystallization nuclei of sub-boundaries (a) and twins (b).

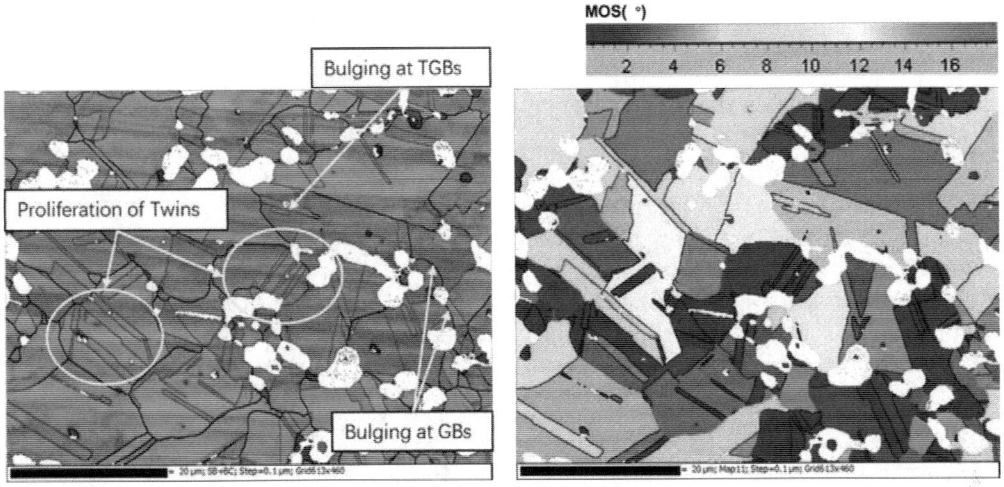

Figure 8. EBSD micrograph show recrystallization nuclei of bulging at TGBs and GBs.

In the process of thermal deformation of Stellite 6B alloy, whether at lower temperature or higher temperature, recrystallization first nucleates at the dislocation plug near the carbide. When the critical strain is reached, a large number of recrystallization cores are rapidly formed in the form of subgrains rotation or subgrains coalescence. Dynamic recrystallization is achieved by developing from subgrains coalescence into recrystallized grains at higher temperature. With the increase of deformation, recrystallized grains grow significantly. At lower temperature, dynamic recrystallization nucleates in the form of subgrains rotation and develops into recrystallized grains.

4. Conclusions

Hot compression test in the temperature range of 1000-1200 ℃, and strain rate range of 0.01-10 s^{-1} of Stellite 6B Co-based alloy has been conducted. The following conclusions can be drawn from of the results of this investigation:

(1) The dynamic crystallization was beginning to occur in all processing conditions when the strain was closed to 0.1. The recrystallization was preferentially nucleated at the nearby regime of a series-shaped carbides, the dynamic recovery process is inhibited, and the softening effect by dynamic recrystallization is more significant.

(2) With the increase of deformation, recrystallization nucleates at the bending of the original grain boundaries and twin boundaries. Twins play an important role in the dynamic recrystallization of Stellite 6B alloy. The recrystallization grain size of the alloy is determined by the disappearance and proliferation of twins.

References

[1] Yao M X, Wu J B C, Xu W and Liu R 2005 Metallographic study and wear resistance of a high-C wrought Co-based alloy Stellite 706K Mater *Sci. Eng. A* **407** 291

[2] Malayoglu U and Neville A 2005 Mo and W as alloying elements in Co-based alloys--their effects on erosion–corrosion resistance *Wear* **259** 219

[3] Poliak E I and Jonas J J 1996 A one parameter approach to determining the critical conditions for the initiation of dynamic recrystallization *Acta Materialia* **44** 127

[4] Gourdet S and Montheillet F 2003 A model of continuous dynamic recrystallization *Acta Materialia* **51** 2685

[5] Lee W S, Liu C Y and Sun T N 2004 Deformation behavior of Inconel 690 super alloy evaluated by impact test *Journal of Materials Processing Tech* **153-154** 219

[6] Poelt P, Sommitsch C, Mitsche S, and Walter M 2006 Dynamic recrystallization of Ni-base alloys-Experimental results and comparisons with simulations *Mater. Sci. Eng. A* **420** 306

[7] Cai D, Xiong L, Liu W, Sun G, and Yao M 2007 Development of processing maps for a Ni-based superalloy *Materials Characterization* **58** 941
[8] Liu J, Cui Z, and Ruan L 2011 A new kinetics model of dynamic recrystallization for magnesium alloy AZ31B *Mater. Sci. Eng. A* **529** 300
[9] Amiri A, Bruschi S, Sadeghi M H, and Bariani P 2013 Investigation on hot deformation behavior of Waspaloy *Mater. Sci. Eng. A* **562** 77
[10] Azarbarmas M, Aghaie-Khafri M, Cabrera J M and Calvo J 2016 Dynamic recrystallization mechanisms and twining evolution during hot deformation of Inconel 718 *Mater. Sci. Eng. A* **678** 137
[11] Gujrati R, Gupta C, Jha J S, Mishra S, and Alankar A 2019 Understanding activation energy of dynamic recrystallization in Inconel 718 *Mater. Sci. Eng. A* **744** 638
[12] Wang X, Br€uunger, E, and Gottstein G 2002 The role of twinning during dynamic recrystallization in alloy 800H *Scripta Materialia* **46** 875

Structural changes of 12Cr1MoVG steel during high-temperature creep

Jinping Pan[1,5], Bo Xu[2,5], Xinwei Zhu[1], Xuefen Yu[3,6], Lianjiang Tan[4,6]

[1] Jiaxing Special Equipment Inspection Institute, Jiaxing 314051, China;
[2] Ningbo Special Equipment Inspection and Research Institute, Ningbo 315048, China;
[3] Zhejiang Academy of Special Equipment Science, Hangzhou 310020, China;
[4] School of Materials Science and Engineering, Shanghai Institute of Technology, Shanghai 201418, China

[5]These authors contributed equally to the paper.
[6]Email: 1046794280@qq.com; tanlianjiang@126.com

Abstract. Heat-resistant steels are key metal materials for high-temperature and high-pressure boilers of ultra-super critical (USC) power plants. A lot of steels have been designed for heat resistance purposes, among which 12Cr1MoVG steel is a good alternative but has not been studied intensively for its high-temperature creep properties. In the current work, the 12Cr1MoVG steel was subjected to long-term creep at different temperatures. The 20000 h creep rupture stress showed obvious decrease with increasing temperature in the data range. The structural changes of the 12Cr1MoVG steel during the high-temperature creep tests was investigated by analyzing the cavitation and metallographic structure before and after the creep tests. The results indicated that the structural changes had close relationships with the creep behavior and mechanical properties of the steel. This work gives insights into the high-temperature creep study of heat-resistant steels.

1. Introduction

Steels are the most useful metal materials in human life [1,2]. Ultra-super critical (USC) power plants have been developing intensively due to the urgent demands for clean energy and environment protection. Construction of USC power plants requires materials with excellent high-temperature creep strength and resistance to hot corrosion [3]. In general, a long creep rupture life under a stress of over 100 MPa at the temperatures of ~600 ℃ are required [4]. Varieties of heat-resistant steels have been studied with regard to creeping at high temperatures [5-7]. 12Cr1MoVG steel is a preferable material for fabricating large water wall panel used in boilers. Since safe and economical operations of power plants are crucial, structural changes and cavitation of the heat-resistant steels used in boilers have become a major concern [8-10]. Nevertheless, systematic study for the long-time creep of 12Cr1MoVG steel at high temperatures is insufficient. In the current work, creep tests at high temperatures were performed for 12Cr1MoVG steel. The creep cavitation was investigated to unveil the cavities evolution in the 12Cr1MoVG steel over long-term period of high-temperature creep. Besides, metallographic structure before and after the long-term creep was compared.

2. Experimental

2.1. Materials
The samples of 12Cr1MoVG steel in this work were provided by a USC power plant. The elementary composition was determined by an Optical Emission Spectrometer (ARL 4460, Thermo Scientific). The elementary composition is shown in Table 1.

Table 1. Chemical composition of 12Cr1MoVG steel (at.%).

C	Si	Mn	Cr	Mo	V	P	S
0.109	0.252	0.451	1.110	0.2859	0.186	0.022	0.013

2.2. Creep tests
Tensile creep tests for 12Cr1MoVG samples were carried out at selected temperatures on a high-temperature electronic creep and rupture testing machine (GWT2105, MTS) with constant loads. The samples were processed into rectangular specimen (50*15*3 mm), and two samples were measured in each test.

2.3. Metallographic analysis
12Cr1MoVG steel specimens were processed using a metallographic mosaic machine (CitoPress-10, Struers) to obtain samples with appropriate sizes and shape. Then the samples were processed by an automatic polish-grinding machine (Tegramin-30, Struers) and were etched with 4 % nitric acid in ethanol. The resultant samples were washed thoroughly with ethanol and were observed by an inverted optical microscopy (GX51, Olympus).

3. Results and discussion

3.1. Creep rupture strength
Through tensile creep tests at varied temperatures, the creep rupture strength at 20000 h of 12Cr1MoVG was measured. Figure 1 shows the plot of creep rupture strength at 20000 h versus temperature. As the temperature increased from 450 °C to 625 °C, the creep rupture strength at 20000 h decreased from 130 to 20 MPa, demonstrating the significant effects of temperature on the creep rupture life of the 12Cr1MoVG steel. In the temperature range characteristic of the boilers in operation, the creep behavior of the 12Cr1MoVG steel is sensitive to the temperature changes. The data of creep rupture strength help to evaluate the work conditions of the 12Cr1MoVG steel and predict the service life of the 12Cr1MoVG steel under the work conditions.

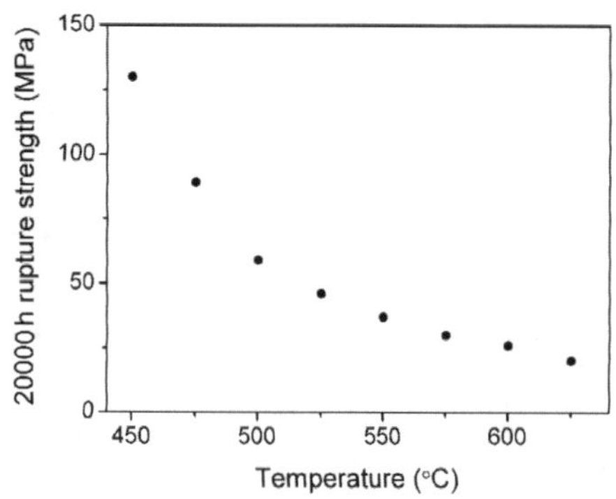

Figure 1. Changes of creep rupture strength at 20000 h with temperature for the 12Cr1MoVG steel.

3.2. *Structural changes*

Materials have various creep damage mechanisms. Nevertheless, some characteristics are common, such as growth and coalescence of cavities into microcracks, and then into macroscopic cracks that become detectable by conventional non-destructive technique [11]. We measured the cavity density in the 12Cr1MoVG steel within a certain area. The results in Figure 2a indicate that the cavity density decreased from around 3700 to around 100 per square millimeters when the distance from the notch tip increased to 4 mm. This suggests that cavities start to appear in vicinity of the edge of a sample. Thereafter, more cavities are produced, both in the vicinity of the edge and in the area away from the edge. As indicated in Figure 2b, the average cavity size over the area measured fluctuated within a certain range from around 2.8 to 4.4 μm. These cavities gradually grew to larger sizes, coalescing into larger cavities or cracks, and finally formed macroscopic cracks that would lead to rupture of the steel. It was the formation and evolution of cavities in the 12Cr1MoVG steel that resulted in deteriorated mechanical properties during the long-term creep at a high temperature. Compared with 15CrMoG steel reported in our previous work [12], the density of cavities in the 12Cr1MoVG steel was lower under similar creep conditions. The average cavity diameter was close to that of the 15CrMoG steel. These results indicate that the 12Cr1MoVG steel has higher creep resistance at its service temperatures.

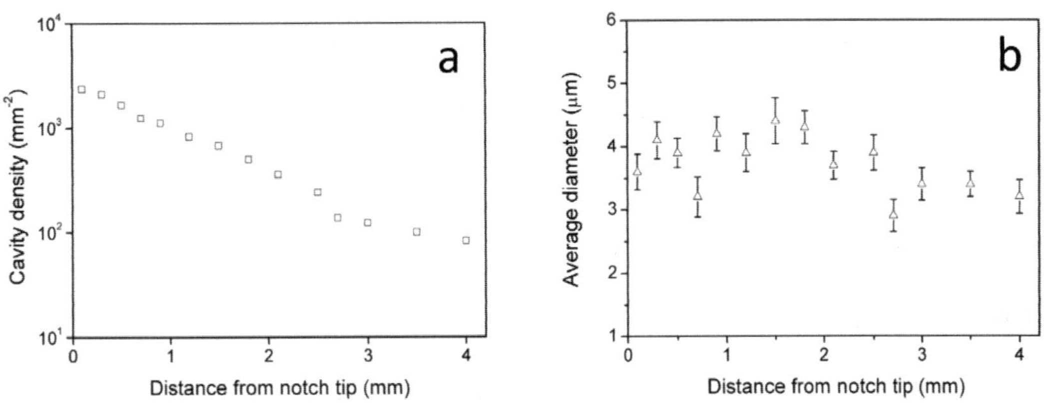

Figure 2. (a) Cavity density and (b) average diameter of cavities in the 12Cr1MoVG steel with a size of 4.5×3 mm^2. The quantity of cavities was measured along the tensile direction. The data in plot b are presented as the average ± standard deviation.

In order to further investigate the changes in microstructure of the 12Cr1MoVG steel subjected to the long-term creep, metallographic images of an untreated sample and a sample experiencing creep at 600 °C over 20000 h were obtained, as shown in Figure 3. For the untreated sample (Figure 3a), the dark regions denote unspheroidized pearlite, where the carbide exists in lamellar form. For the sample after long-term creep (Figure 3b), the light regions denote the pearlite and ferrite. The carbide was distributed at the grain boundaries, suggesting that pearlite spheroidization took place in the sample subjected to the creep tests at the selected temperature.

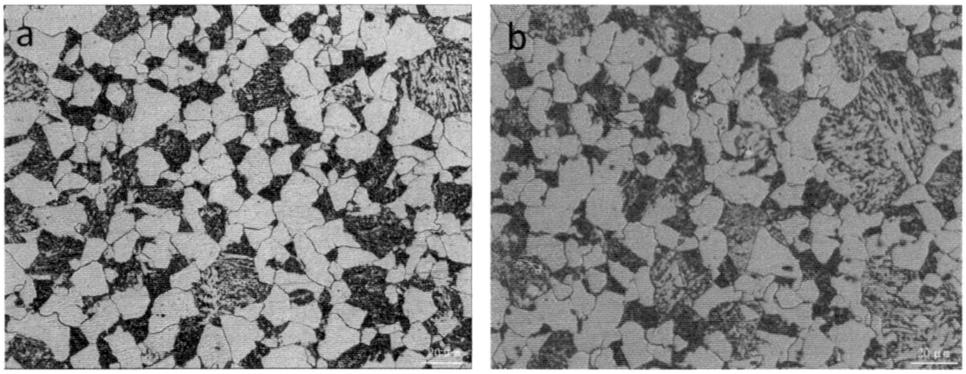

Figure 3. Metallographic images of (a) an untreated 12Cr1MoVG sample and (b) a sample subjected to creep at 600 °C over a period of 20000 h.

4. Conclusions

To sum up, the 12Cr1MoVG steel has been subjected to long-term creep tests at selected temperatures, based on which the creep rupture strength at 20000 h was obtained. According to the long-term creep rupture data, appropriate work conditions for the 12Cr1MoVG steel as a key material for boilers can be determined. The structural changes of the 12Cr1MoVG steel during the long-term creep were studied by analyzing the evolution of cavities in the steel and observing the metallographic changes. The cavitation evolution including migration and growth of cavities with the time account for the reduced mechanical performance of the steel. Subjected to the creep, the 12Cr1MoVG steel showed the phenomenon of pearlite spheroidization, as reflected by the distribution of carbide at the grain boundaries. The findings of this work provide useful information on the creep properties at high temperatures of 12Cr1MoVG steel and can guide the safe service of the boilers using 12Cr1MoVG steel as the material.

Acknowledgment

This work is financially supported by the Jiaxing Science and Technology Project (2019AD32016).

References

[1] Roy K, Ting T C H, Lau H H, Lim J B P 2019 Experimental and numerical investigations on the axial capacity of cold-formed steel built-up box sections. *J. Constr. Steel. Res.* **160** 411-427
[2] Ting T C H, Roy K, Lau H H, Lim J B P 2018 Effect of screw spacing on behavior of axially loaded back-to-back cold-formed steel built-up channel sections. *Adv. Struct. Eng.* **21** 474-487
[3] Aghajani A, Somsen Ch and Eggeler G 2009 On the effect of long-term creep on the microstructure of a 12% chromium tempered martensite ferritic steel. *Acta Mater.* **57** 5093-5106
[4] Rojasa D, Garciab J, Prata O, Carrascoc C, Sauthoffa G, Kaysser-Pyzalla A R 2010 Design and characterization of microstructure evolution during creep of 12% Cr heat resistant steels. *Mater. Sci. Eng. A* **527** 3864-3876
[5] Guo L, Sui Y-h, Zhang X-Z (2018) High-temperature creep constitutional model of Q460E steel and effect of creep on bulging deformation of continuous casting slab. *J. Iron Steel Res. Int.* **25** 1123-1130
[6] Pan J P, Tu S H, Sun G L, Zhu X W, Tan L J, Hu B 2018 High-temperature creep properties and life predictions for T91 and T92 steels. *IOP Conference Series: Materials Science and Engineering* **292** 012098
[7] Peng Y-Q, Chen T-C, Chung T-J, Jeng S-L, Huang R-T, Tsay L-W 2017 Creep Rupture of the

simulated HAZ of T92 steel compared to that of a T91 steel. Materials **8** 139
[8] Fedoseeva A, Dudova N, Kaibyshev R 2016 Creep strength breakdown and microstructure evolution in a 3%Co modified P92 steel. *Mater. Sci. Eng. A.* **654** 1-12
[9] Vallant J C, Vandenberghe B, Hahn B, Heuser H and Jochum C 2008 T/P23, 24, 911 and 92: New grades for advanced coal-fired power plants-properties and experience. *Int. J. Pres. Ves. Pip.* **85** 38-46
[10] Cerjak H, Hofer P and Schaffernak B 1999 The influence of microstructural aspects on the service behaviour of advanced power plant steels. *ISIJ Int.* **39** 874-888
[11] Auerkari P, Holmström S, Veivo J and Salonen J 2007 Creep damage and expected creep life for welded 9-11% Cr steels. *Int. J. Pres. Ves. Pip.* **84** 69-74
[12] Pan J, Tu S, Zhu X, Tan L 2019 Creep behaviour and cavitation evolution of 15CrMoG steel at high temperatures. *Adv. Mech. Eng.* **11** 1-7

IOP Conf. Series: Materials Science and Engineering 1040 (2021) 012014 doi:10.1088/1757-899X/1040/1/012014

Research on ball milling process of Mn powder for ceramic metal welding

Baijiang An[1]

Shaanxi Baoguang Vacuum Electric Device Co., Ltd, Baoji 721016, Shaanxi, China

[1]Email: anbaijiang@126.com

Abstract. The main content of this experiment is to study the optimal ball milling process of Mn powder by controlling the factors such as ball milling ratio, ball milling time and grinding ball size. In the experiment, the Mn powder was ground by wet ball milling method. Ethanol was used as the medium to set different powder: ball: ethanol ratio. The microstructures of Mn were studied by grinding for 6 h, 18 h and 24 h at 400 r/min. The XRD, Scanning Electron Microscope and particle size analyzer were used to analyze the phase, microstructure and particle size of the powder. The results show that the ratio of the crushing rate of material to the time of ball milling are not related. In the initial stage of ball milling, Mn powder was rapidly crushed and refined, but the particle size gradually stabilized after 18 hours. The grinding ball with smaller diameter is slower to crush, but the particle size of Mn powder is more uniform. When the quality of powder: ball: ethanol =1:1.5:1.5, the obtained Mn powder has the finest particle size and uniform particles, which is the best grinding condition. After ball milling, the average particle size of Mn powder is 2.04 μm.

1. Introduction

Alumina ceramics have widely used in electric vacuum devices because of its non-conductive, high hardness and high temperature resistance. In the process of ceramic and metal welding, it is a common process operation to coat the surface of alumina ceramics with molybdenum, manganese and other conductive metal powder [1] to achieve the purpose of modification. The product not only has high temperature resistance, but also has good electrical conductivity, which has a high application prospect in aerospace, energy technology and other industries[2-4].

The activated Mo-Mn method is mainly used in the welding of ceramics and metals. In the process of ceramic metallization, Mn is a kind of very important raw material and is also one of the frontier fields of manganese material research. The particle size of Mn powder plays an important role in ceramic metal sealing. However, high energy mechanical ball milling [5,6] is a common method for powder grinding at room temperature. There are many factors affecting ball milling, such as raw material properties, ball milling strength, ball milling atmosphere, ball milling ratio, ball milling time and ball milling temperature[7-9]. In this paper, the method of obtaining manganese powder with smaller particle size by high-energy ball milling is also a more convenient and clean method than hot processing. The finer the powder, the better the sintering, and the higher the strength of ceramic metal sealing. According to the experimental data, the grinding ball parameters corresponding to the optimal microstructure and sintering strength can be obtained to promote the improvement of Mn powder processing technology and the progress of related industries.

2. Experimental part

Mn powder used in the experiment is analytically pure, purchased from Shaanxi Wenzheng Electronic Technology Co., Ltd., with the average particle size of 9.35 μm and the particle size was analyzed by LS-POP(VI) laser particle size analyzer, as shown in Figure 1.

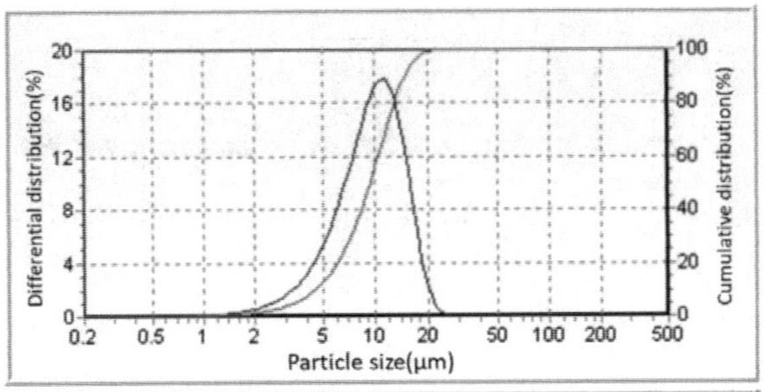

D(4,3) :9.58 μm	D50 :9.35 μm	D(3,2) : 7.40 μm	S.S.A. :0.81 sq.m/c.c.
D10 :4.54 μm	D25 :6.67 μm	D75 :12.19 μm	D90 :14.81 μm

Figure 1. Particle size distribution diagram and characteristic parameters of Mn particle size.

The ball milling was carried out under normal temperature and pressure. The aluminum oxide ball mill pot was used with a volume of 300 mL. The average diameter of the big ball and the small ball was 8.38 mm and 5.90 mm respectively. In the experiment, ultra-fine Mn powder, agate balls with different particle sizes and absolute ethanol were added into the ceramic ball mill according to different proportions. The samples were milled by planetary high-energy ball mill. The samples were dried for 6 h, 18 h and 24 h in a constant temperature blast drying oven, and then they were passed through Japanese physics Dmax Ultima. The phase structure (XRD) patterns of the samples were obtained by measuring the X-ray diffraction of the samples with type I X-ray diffractometer. The microstructure of the samples was observed by quanta 250EFG scanning electron microscope at 20 kV.

3. Results and discussion

3.1. Effect of ball milling time

The main mechanism of ball milling is the impact between the high-speed ball and the material to achieve the purpose of crushing materials. At the same time, the length of milling time is bound to affect the quality of materials. Figure 2 (a), (b) and (c) are the SEM images of the samples obtained at 6 h, 18 h and 24 h under the ball milling ratio of m_{powder}: m_{ball}: $m_{ethanol}$ = 1:1.5:1.5. The magnification ratio of the three graphs is 3000 times. It can be seen from the figure that the particle size of Mn powder decreases with the time going on, and tends to be stable after 18 h. This is because in the initial stage of ball milling, the Mn powder particles are larger, the momentum generated is also larger, and a lot of energy is released when colliding with the grinding ball[10], which promotes the crushing of Mn powder. After that, with the gradual reduction of particles, the chance of ball catching powder becomes less, and the surface free energy per unit volume increases gradually, which makes it more difficult to further refine the materials. At the same time, the crushing and welding of metal particles are gradually balanced in the process of ball milling, which is also the reason why the crushing rate and time of materials change irregularly in the process of ball milling.

Figure 2. SEM images of Mn powder under different milling time when the ball milling ratio is m_{powder}: m_{ball}: $m_{ethanol}$ = 1:1.5:1.5 : (a)6 h,(b)18 h and (c)24 h.

3.2. Effect of ball milling ratio

In ball milling, if there are too many grinding balls, the powder will collide, and if the alcohol is too much, the fluid will be formed. Therefore, it is necessary to set the appropriate ball milling ratio. In Figure 3, the SEM images of samples with different ratios of Mn powder and grinding ball running for 24 h are shown under 3000 times magnification. It can be seen that the particle size of the powder in Figure 3(d)(f)(h) are smaller than that of the others, while the milling effect of (f) is better than that of (d) and (h). Table 1 is obtained after the statistics of the average diameter of Mn powder in the following proportion images.

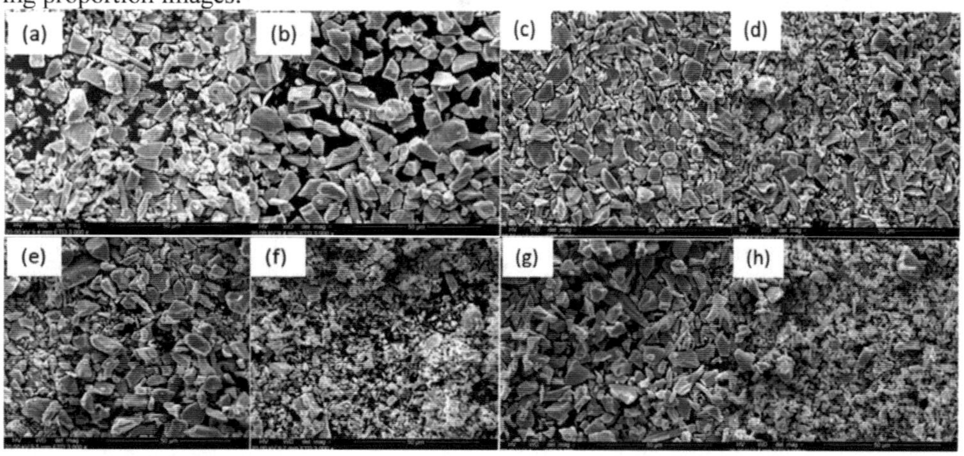

Figure 3. SEM images of Mn powder milled for 24 h under different ball milling ratio (m_{powder}: m_{ball}: $m_{ethanol}$): (a) 2:1:2 (b) 1:1:1 (c) 2:1:1 (d) 2:1:1.5 (e) 1:1:1.5 (f) 1:1.5:1.5 (g) 1.5:1:1.5 (h) 1.5:1.5:1.

Table 1. Average diameter of Mn powder under different ball milling ratio.

Sample	Ball milling ratio / (m_{powder}: m_{ball}: $m_{ethanol}$)	Average diameter D50/μm
1-1	2:1:2	6.13
2-1	1:1:1	6.45
3-1	2:1:1	4.00
4-1	2:1:1.5	3.59
5-1	1:1:1.5	5.71
6-1	1:1.5:1.5	2.04
7-1	1.5:1:1.5	4.59
8-1	1.5:1.5:1	3.19

Compared with Figure 3 and Table 1, it can be seen that only when the proportion of ball and ethanol is higher than that of Mn powder, can Mn powder with smaller particle size be obtained. At this time, on the one hand, the chance of collision between Mn powder and grinding ball is increased, while ethanol can fully contact with Mn powder, which reduces the welding between particles, thus making the result of ball milling more ideal and the particle size of Mn powder smaller. The best milling ratio should be m_{powder}: m_{ball}: $m_{ethanol}$ = 1:1.5:1.5, the particle size analysis diagram and characteristic parameters are shown in Figure 4.

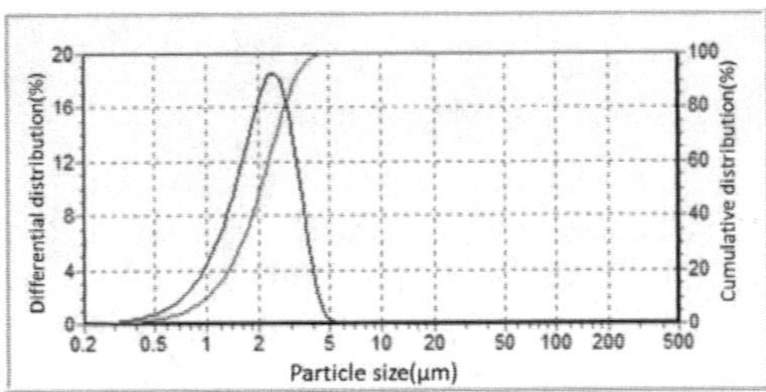

D(4,3) : 2.08 μm	D50 : 2.04 μm	D(3,2) : 1.64 μm	S.S.A. : 3.67 sq.m/c.c.
D10 : 1.01 μm	D25 : 1.47 μm	D75 : 2.64 μm	D90 : 3.19 μm

Figure 4. Particle size distribution diagram and characteristic parameters of particle size with m_{powder}: m_{ball}: $m_{ethanol}$ = 1:1.5:1.5.

3.3. Effect of ball size

The results of ball milling are closely related to the properties of grinding balls. In the experiment, agate ball is a natural mineral, mainly because of its high purity, high hardness, can reduce the pollution of materials, and good wear resistance, can improve the rate of ultrafine grinding. Figure 4 is obtained by controlling different ball size. Figure 5 (a), (b) and (c) are the SEM images of samples milled with 8.38 mm diameter agate ball (hereinafter referred to as large grinding ball) with the mass ratio of m_{powder}: m_{ball}: $m_{ethanol}$ =1:1.5:1.5 ball milling for 6 h, 18 h and 24 h; and (d), (e) and (f) are the agate balls with a diameter of 5.90 mm (hereinafter referred to as small grinding balls) with the mass ratio of m_{powder}: m_{ball}: $m_{ethanol}$ =1:1.5:1.5 ball milling for 6 h, 18 h and 24 h.

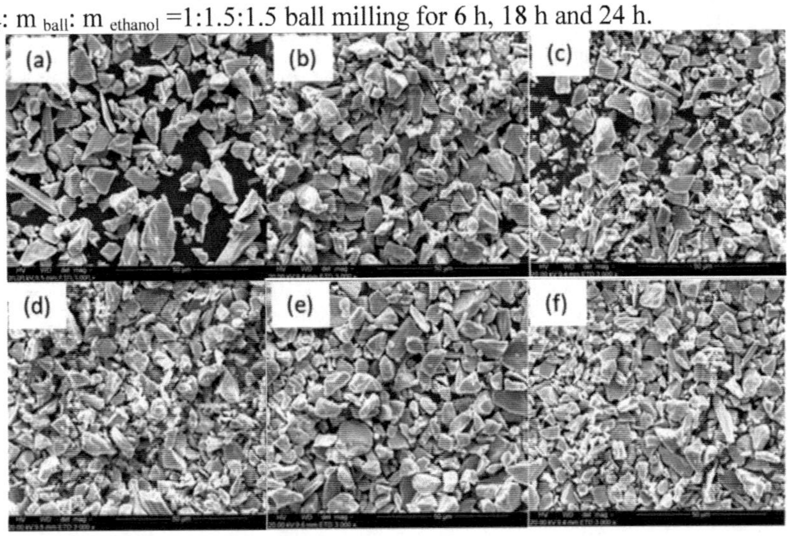

Figure 5. SEM images of Mn powder with different ball size under the same proportion: (a) Large grinding ball 6 h, (b) large grinding ball 18 h ,(c) large grinding ball 24 h, (d) Small grinding ball 6 h, (e) small grinding ball 18 h ,(f) small grinding ball 24 h.

From Figure 5, it can be seen that the momentum generated is naturally large and the material can be quickly crushed in the initial stage of ball milling due to the large mass of the ball mill. However, due to the small number of large grinding balls under the same mass, the chances of collision with Mn powder are less, so the final particles obtained are larger than those of small ones. In the process of ball milling, the volume of agate ball accounts for about 1/2 of the volume of ball mill pot. Due to the larger area of the small ball and the more Mn powder carried by the ball mill than the large ball under the same mass, there are more chances of collision in the process of ball milling, the material is crushed more fully, and the final Mn powder particles are smaller.

4. Conclusion

In this paper, through controlling the factors of ball milling proportion, ball milling time and ball size, powders under different ball milling conditions were obtained. By comparing the particle size and micro morphology of the materials under different factors, the best ball milling process of Mn powder can be obtained. There is no fixed proportion relationship between grinding rate and milling time. At the beginning of ball milling, Mn powder was crushed and refined rapidly, but the particle size gradually became stable after 18 h. The grinding ball with larger diameter grinds faster, but the particle size distribution of Mn powder is uneven. The optimum mass ratio of m_{powder}: m_{ball}: $m_{ethanol}$ = 1:1.5:1.5, and the average particle size is 2.04 μm.

Acknowledgements

This work was financially supported by National Defense Science and Technology Key Laboratory Fund Project (NO. 6142605200103).

References

[1] Wang L, Kang WT, Gao PZ, Liu XP 2020 *Ceram Inl.* **46(6)** 8244
[2] Wu S, Qin Y, Fu D, Fan L, Chen H, Hong H 2020 *Ceram Int.*
[3] Ghosh S, Sengupta A, Pal K. S, Dandapat N, Chakraborty R, Datta S, Basu D 2012 *Metall Mater Trans A.* ***43(3)*** 912
[4] Li L, Guo WG, Yu X, Fu DX 2017 *Ceram Inl.* **43(9)** 6684
[5] Salah N, Habib SS, Khan ZH 2011 *Int J nanomed.* **6** 863
[6] Gusev AI, Kurlov AS 2008 *Nanotechnology.* **19(26)** 265302
[7] Yadav TP, Yadav RM, Singh DP 2012 *Nanosci Nanotech.* **2(3)** 22
[8] Zhang P, Jia D, Yang Z, Duan X, & Zhou Y 2013 *Ceram Int.* **39(2)** 1963
[9] Enqvist E, Ramanenka D, Marques PA, Gracio J, Emami N 2016 *Polym Composite.* **37(4)** 1128
[10] Chen C, Liu Z, Xiang J, Liu S, Zhang Y, Wang H 2012 *Journal of Yanshan University.* **36(2)** 136

Synthesis and characterization of Ag nanoflowers with different morphologies

Huiqin Li[1], Shaotian Yan, Cui Deng, Shumei Dou[1] and Xinyue Ren

College of Chemistry and Chemical Engineer, Baoji University of Arts and Sciences, Baoji 721013, Shaanxi, China

[1]Email: huiqinli@yeah.net (Li H); dsmwxsh@163.com (Dou S)

Abstract. In this paper, the chemical reduction method was used to prepare silver nanoparticles with different morphologies and sizes by controlling the proportion of ascorbic acid and silver nitrate at room temperature. In the experiment, the molar ratio of ascorbic acid and silver nitrate was mainly 1:1, 1:2, 1:4, 1:8, 2:1, 4:1 and 8:1 to prepare nano silver. The morphology of silver nanoparticles was observed by scanning electron microscopy (SEM) and the phase content was determined by X-ray diffraction (XRD). The results show that the obtained material is pure silver nanoparticles, and the silver nanoparticles are spherical and multilevel branched.

1. Introduction

In recent years, noble metal nanoparticles have been widely used in photonics [1], microelectronics [2], information storage, catalysts [3-6], energy conversion [7] and other fields. Among all nanoparticles, silver nanoparticles play an important role due to their unique chemical properties and excellent biological activity, and have the most effective surface enhanced Raman scattering (sere) effect in the visible region [8-9]. In the context of the development of information technology, nanotechnology has made groundbreaking achievements in many fields. It is of great theoretical significance and potential value to carry out basic research on nanotechnology [10-11].

Nano silver is a kind of powder with particle size less than 100 nm, which can be synthesized by physical and chemical methods. The chemical method has the advantages of easy control of experimental conditions, easy operation and easy access to products, so it is widely used to synthesize silver nanoparticles with good performance. Compared with traditional metal nanoparticles, nano silver has strong advantages in antibacterial property [12-13], surface activity and catalytic performance, so it is widely used in medicine, catalyst, electronic industry, textile industry, agriculture, detection and other fields. Because of its high antibacterial activity, nano silver has become a new star in the field of nano materials research; moreover, it has the advantages of low price and high performance price ratio, which is more suitable for industrial application than other nano materials.
Nano silver materials are widely used. We can design and prepare nano silver materials with specific morphology and size, so as to improve the utilization efficiency of nano materials. Based on this, this paper aims to synthesize silver nanoparticles with different sizes and morphologies by the molar ratio of ascorbic acid to silver nitrate.

2. Experimental part

2.1. Experimental reagents and instruments

The reagents and instruments used in the test are listed in Table 1 and table 2 respectively.

Table 1. Introduction of reagents.

Name of drug	purity	Manufacturer
$AgNO_3$	99.8%	International Group Chemical Reagent Co., Ltd
L-ascorbic acid (c)	99.7%	International Group Chemical Reagent Co., Ltd

Table 2. Introduction of related instruments

Instrument name	purity	Manufacturer
Analytical balance	BS221S	Beijing saiduolis balance Co., Ltd
Desktop centrifuge	H1650-W	Xiangyi company
Drying oven	GZX-GF101-1-BS-II	Shanghai Yuejin Medical Equipment Co., Ltd
Field emission SEM	Quanta FEG 250	USA,FEI company
X-ray diffractometer	D/MAX-3BX	Japan, RIGAKU

2.2. Experimental steps

10 mL of 1.0 mol / L silver nitrate solution was mixed with 10 mL of 1.0 mol / L ascorbic acid solution at room temperature and stirred. When the ascorbic acid solution was added into silver nitrate solution, the solution turned to gray black turbid liquid, and then turned to gray yellow, and a large amount of precipitation was formed in a few minutes. Then the molar ratio of ascorbic acid and silver nitrate was changed. In this experiment, seven groups were made, which were 1:1, 2:1, 4:1, 8:1, 1:2, 1:4, 1:8. The precipitates were centrifuged and cleaned with distilled water, and centrifuged at 8000 r / min for six minutes. After centrifugation, the precipitates were dispersed in water to prepare for further characterization. The phase of the samples was measured by D / max-3b X-ray diffractometer and the morphology was observed by field emission scanning electron microscope.

3. Results and discussion

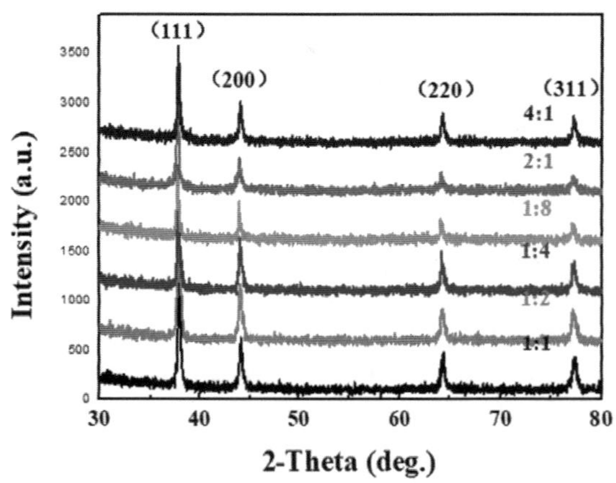

Figure 1. XRD pattern of silver nanoparticles.

Figure 1 shows the XRD pattern of silver nanoparticles. Scanning angle 2 θ was performed from 30° to 80 °. The diffraction planes shown are (111), (200), (220), (311), which are in good agreement with the standard PDF card (# 65-3107). The results show that the material is pure silver nanoparticles.

Figure 2. (a), (b), (c), (d), (e), (f) and (g) are respectively [Vc] : [AgNO3] is 1:1; 2:1.;4:1; 8:1;1:2; 1:4 SEM image.

Figure 2 shows the SEM images obtained when the molar ratio of ascorbic acid to silver nitrate is different. It can be seen from Figure 1 that the silver microcrystals obtained by changing the molar ratio of ascorbic acid to silver nitrate (MR) are different. When the molar ratio of ascorbic acid to silver nitrate is reduced from 1:2 to 1:4 and 1:8, it can be seen that the size of silver nanoparticles increases significantly and the morphology does not change. However, when the molar ratio of ascorbic acid to silver nitrate increased from 2:1 to 4:1 and 8:1, the number of silver microcrystals with dendritic hierarchical structure decreased.

4. Conclusion

The unique structure and properties of silver nanoparticles have attracted much attention. Ascorbic acid is both a reducing agent and a crystal growth regulator in this experiment. The core idea of this paper is to study the effect of ascorbic acid content on the morphology and size of silver nanoparticles. In this paper, the preparation and characterization of silver nanoparticles are discussed. In this paper, we use ascorbic acid as reducing agent to prepare silver nanoparticles at room temperature. It is known from the experiment that different [Vc]: [$AgNO_3$] molar ratio plays an important role in the structure of silver nanoparticles, and the size and morphology of silver nanoparticles obtained by different ratios are completely different. It can be seen from SEM images that in order to prepare graded silver microcrystals, 2:1 molar ratio of [Vc]: [$AgNO_3$] is the best, and the distribution of silver microcrystals under this ratio is relatively uniform. The XRD results show that the samples are pure silver nanoparticles.

Acknowledgements

This work was financially supported by the National Natural Science Foundation of China (NO.51702006), Shaanxi provincial science and technology planning project (2018JQ2056) , the Education Commission of Shaanxi Province (NO.18JK0053), Innovation project (NO.2019XJ093) and innovative research projects for Postgraduates (NO.YJSCX19YB09) of Baoji University of Arts And Sciences.

References

[1] Wang X, Ruditskiy A, Xia Y 2016 *Nati Sci Rev.***3(4)** 520
[2] Li J, Zhang X, Liu L, Han L 2012 *J. Microelectrome Ch S.* **22(3)** 560
[3] Pan L, Ji Z, Yi X, Zhu X, Chen X, Shang J, Li RW 2015 *Adv Func Mater.* **25(18)** 2677
[4] Zhang J, Wang H, Wang L, Ali S, Wang C, Wang L, Xiao FS 2019 *J. American Chem Soc.* **141(7)** 2975
[5] Fang X, Li J, Ren B, Huang Y, Wang D, Liao Z, Dionysiou DD 2019 *J.Membrane Sci.* **579** 190
[6] Das TK, Bhawal P, Ganguly S, Mondal S, Das NC 2018 *Surf Interfaces.* **13** 79
[7] Fan Y, Hu G, Yu S, Mai L, Xu L 2019 *Sci China Mater.* **62(3)** 325
[8] Cheng ZQ, Shi HQ, Yu P, Liu ZM 2018 *Acta Phys Sin-Ch Ed.* **67(19)** 197302
[9] Wang J, Yi G 2019 *Nanoscale Res Lett.* **14(1)** 1
[10] Xu Y, Li S, Yue X, Lu W 2018 *BioResources.* **13(1)** 2150
[11] Liu W, An R, Wang C, Zheng Z, Tian Y, Xu R, Wang Z 2018 *Micromachines.* **9(7)** 346
[12] Wang Q, Qian Z, Liu B, Liu J, Zhang L, Xu J 2019 *J Biomat Sci-Polym E.* **30(6)** 462
[13] Yun Q, L C, Li R, Liu G, Zhang Y, Tang X, Wang J, Qin Y 2018 *Int J Nanomed.* **13** 3311

Potential of limestone powder to improve the stabilization of sulfate-contained saline soil

Ayazhan Bazarbekova[1,2], Chang-Seon Shon[1], Aizhan Kissambinova[1], Jong Ryeol Kim[1], Dichuan Zhang[1], and Sung-Woo Moon[1]

[1]Department of Civil and Environmental Engineering, School of Engineering and Digital Sciences, Nazarbayev University, Nur-Sultan, Kazakhstan, 010000

[2]Email: ayazhan.bazarbekova@nu.edu.kz

Abstract. The geotechnical properties of soils differ from area to area, depending on the origin of soil, environmental conditions of a region, and soil treatment processes. As a subgrade material in pavement construction, however, the soil is required to meet specific standards for its engineering properties, such as plasticity, deformability, strength, and durability parameters. For this purpose, soil undergoes a stabilization process, in which it is treated with chemical additives. This paper studies the potential of limestone powder as a stabilizer in combination with cement in the stabilization of sulfate-contained saline soil. For this purpose, silty sand (A-2-4(0) according to AASHTO soil classification system) containing high sulfate and sodium contents was stabilized with 6% and 8% cement by dry weight, 2% and 4% limestone powder at fixed 4% cement content, and 2% limestone powder by dry weight at 6% cement content. Series of tests were performed to determine the improvement of geotechnical properties of these mixtures, such as unconfined compressive strength, Atterberg limits, three-dimensional (3-D) swelling, and dielectric constant. Experimental results show that the addition of limestone powder to the cement-treated saline soil decreases soil plasticity, increases maximum dry density, improves strength parameter, reduces volumetric swelling and moisture susceptibily of soil.

1. Introduction

Weak and soft subgrade soils often result in poor performance and a short lifetime of road pavements. For example, excessive heave occurs in pavements constructed on saline soils containing medium to high sulfate level due to high localized stresses and non-uniform movement of structures in soil caused by crystallization pressure, which is created by salt whiskers in soil. In order to improve the poor quality of soils and meet the desired end performance criteria in pavement construction, soil undergoes a stabilization process, in which it is treated with chemical additives, usually, lime, Portland cement, and calcium chloride [1, 2].

In recent years, some researchers have evaluated the effect of limestone powder as an alternative binder for soil stabilization. It has been reported that limestone powder, when added to fine-grained soils, improves the bearing capacity and strength parameters of soil [3]. Moreover, stabilization by limestone powder allows for the reduction of deformability of weak clayey soil, thus, leading to reducing the thickness of road pavement constructed on the stabilized soil [4, 5].

However, recent studies on stabilization of soil using limestone powder have most likely focused on the evaluation of physical and mechanical properties, not durability parameters. Moreover, little data are available on the effect of limestone powder on the stabilization of sulfate-contained saline soil.

Therefore, this study aims to evaluate the improvement of geotechnical properties and durability parameters of sulfate bearing saline soil treated with limestone powder and its combination with cement.

2. Materials, mixtures, and methods

2.1. Materials

2.1.1. Soil. The soil tested in the current study was natural clayey, gravel, sandy, high sulfate-contained saline soil collected from West Kazakhstan as shown in Table 1. Basic geotechnical properties including soil classification, Atterberg limits (liquid limit (LL), plastic limit (PL), plasticity index (PI)), and moisture content-dry density relationship were determined according to the AASHTO soil classification system [6], ASTM D4318-17 [7], and ASTM D698-12e2 [8], respectively. The chemical properties of soil such as dissolved cations and anions contents were determined by the ion chromatography (IC) method.

Table 1. Geotechnical and chemical properties of tested soil.

Property	Value	Property		Value
AASHTO Classification	A-2-4(0)	Liquid limit (%)		19.16
Optimum moisture content (%)	10.8	Plastic limit (%)		16.67
Maximum dry density (kg/m^3)	1941	Plasticity index (%)		2.49
Determination of dissolved cations and anions (mg/L)				
Cations	Calcium	Sodium	Potassium	Magnesium
	693	29,056	437	7229
Anions	Sulfate		Chloride	
	22,495		52,402	

2.1.2. Additives. Limestone powder was obtained by crushing locally collected limestones using a jaw crusher and then, grounding the material using a ball mill. Both limestone powder and cement were sieved through #200 sieve (75 μm) and used as stabilizers at different contents, which were determined using the Edges-Grim method described in TxDOT (Tex-121-E [9]) specification. According to this testing method that determines the recommended percentage of a binder to be used for soil stabilization, the minimum limestone powder and cement contents are 2% and 4%, respectively.

2.2. Mixtures

A total of 6 mixtures, including a control sample with no stabilizer, were designed. As stated previously, the minimum limestone powder and cement percentages needed for soil mixing were 2% and 4%, respectively. Since this study focuses on the potential of limestone powder to benefit the stabilization of sulfate-bearing saline soil, the effect of cement was compared to the combined effect of cement and limestone powder. Thus, the following 6 mixtures were designed and tested: soil + 6% cement, soil + 8% cement, soil + 4% cement + 2% limestone powder, soil + 4% cement + 4% limestone powder, soil + 6% cement + 2% limestone powder.

2.3. Methods

The experimental part of the study can be categorized into material and sample preparation, determination of geotechnical properties, and evaluation of the durability of soil-cement-limestone powder mixtures. Geotechnical properties determined in this study include Atterberg limits, optimum moisture content, maximum dry density, and 7- and 28-day unconfined compressive strength (UCS). Durability assessment was obtained by measuring 3-dimensional (3-D) swelling and dielectric constant of stabilized mixtures. While the dielectric constant test assesses the moisture susceptibility of soil mixture, the 3-D swelling test assesses the volumetric expansion of the sample caused by the ettringite formation when exposed to prolonged capillary soak conditions.

Soil samples were mixed thoroughly with different percentages of cement and limestone powder at their corresponding moisture contents and compacted with standard proctor compaction energy. The Atterberg limits corresponding to LL, PL, and PI were measured for each soil mixture. The moisture content-dry density relationship (M-D curve) of the natural soil was obtained as described in ASTM D698-12e2 [8], and then OMC for the stabilized mixtures was calculated according to the soil-cement testing proposed by TxDOT (Tex-120-E [10]). The following Equation (1) allows to calculate nearer OMC for stabilized soil sample without running a new moisture content-dry-density relationship for each cement content:

$$\%\ molding\ water = \%\ OMC\ from\ M - D\ curve + 0.25 \times (\%\ cement\ increase) \quad (1)$$

Maximum dry densities of soil-cement-limestone powder mixtures were calculated as shown in Equation (2). The moist density of specimen is a density of soil stabilized with cement or its combination with limestone powder, mixed at the corresponding optimum moisture content, and compacted with standard Proctor compaction energy.

$$Dry\ density = \frac{Moist\ density}{1 + \frac{OMC}{100}} \quad (2)$$

$$Moist\ density = \frac{Wet\ mass}{Volume} \quad (3)$$

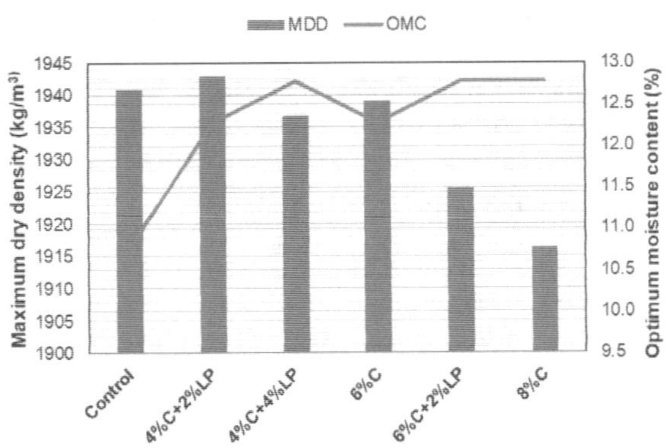

Figure 1. Optimum moisture content and maximum dry density.

The 7- and 28-day UCS of the natural soil and stabilized mixtures were measured as described in ASTM D2166 [11]. Samples of 50 mm diameter and 100 mm height were compressed at a loading rate equal to 1 mm/min under unconfined conditions, and the maximum load that the cylindrical samples could withstand was determined as UCS. 3-D swelling and dielectric constant tests were performed by placing 101.5 mm diameter and 114.3 mm height sample covered with membrane and porous stones in a container, pouring a little amount of water, letting it stay for a particular period, and measuring volumetric expansion and dielectric constant of the tested sample.

3. Test results and discussions

3.1. Atterberg limits

According to the Atterberg limits of the designed mixtures, cement and its combination with limestone powder, when used in soil stabilization, reduce the plasticity of the soil. As presented in Figure 2, an 18% decrease in LL corresponding to the addition of 8% cement can be compared to 16% and 17% decrease in LL of soil mixtures with 4% cement content combined with 2% and 4% limestone powder, respectively. It shows that the combined effect of cement and limestone powder is as strong as the effect of cement used in a higher percentage. Moreover, the addition of 6% cement combined with 2% limestone powder results in a significant drop in PI value, particularly 86% reduction, which is almost a similar 91% decrease of PI for 8% cement-treated soil. This again shows that higher cement content can be replaced by lower cement content with the addition of a small amount of limestone powder, as two mixtures have almost the same Atterberg limits.

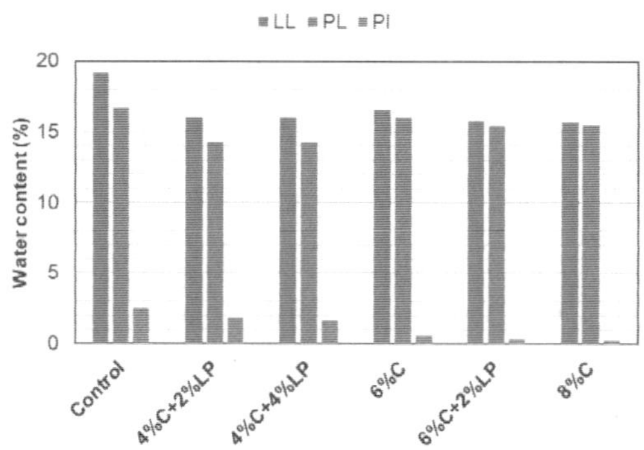

Figure 2. Atterberg limits.

3.2. Optimum moisture content and maximum dry density

The values of optimum moisture content and maximum dry density for the stabilized mixtures were obtained using Equations (1) and (2), respectively, and are illustrated in Figure 1. In general, the stabilization of soil aims to increase maximum dry density while decreasing the moisture content of the soil. However, except for the soil sample treated with a combination of 4% cement and 2% limestone powder, all other mixtures show a decrease in maximum dry density value. The reduced dry densities may be attributed to an increase in resistance against compaction caused by flocculated and agglomerated soil particles formed during cation exchange between ions surrounding soil surface and ions derived from stabilizers [12]. The optimum moisture content increases with an increase in total stabilizer content, as flocculated particles occupy larger spaces and increase the void ratio of soil [12].

3.3. Unconfined compressive strength (UCS)

As shown in Figure 3, the stabilized mixtures have considerably higher unconfined compressive strength compared to the natural soil, which corresponds to the main purpose of soil stabilization. The maximum increase in strength up to 171% was obtained by stabilizing the naturally saline soil with 8% cement. In general, cement content increase correlates with strength improvement, whereas limestone powder content increase from 2% to 4%, at fixed 4% cement content, reduces soil strength. During mixing, the negatively charged surface of sulfate bearing soil attracts positive ions derived from cement and limestone powder (e.g., Ca^{2+}) which replace weak cations from the negatively charged surface and forms flocculated soil particles that contribute to higher surface tension and, as a result, short-term soil strength improvement [1, 13]. Long-term strength improvement, particularly 28-days UCS, is attrubited to the formation of calcium aluminate hydrate (C-A-H) and calcium silicate hydrate (C-S-H), products of cement hydration and pozzolanic rection during soil stabilization woth cement and limestone powder.

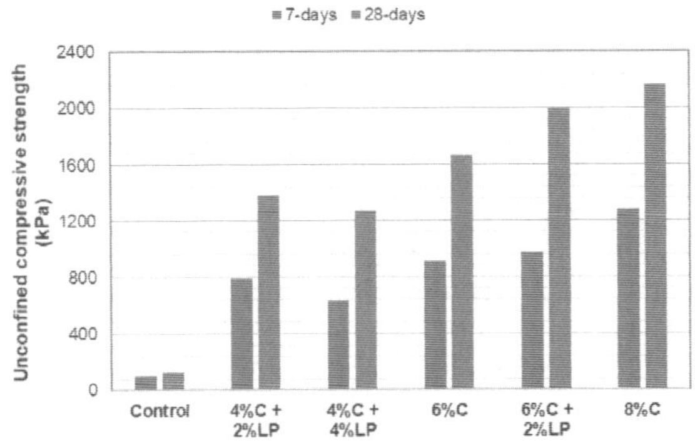

Figure 3. Unconfined compressive strength.

3.4. 3-D swelling

Figure 4 shows the volumetric expansion results over time obtained from the 3-D swelling test for the designed soil mixtures. All soil mixtures experienced rapid volumetric swelling at an early stage. This steep increasing volumetric expansion up to 4-day can be attributed to the moisture capillary suction of the dried sample. That was then followed by steady volumetric expansion at a slower rate for all soil samples. Such behavior may be caused by ettringite minerals formed during the reaction of aluminum, sulfate, calcium, and water in cement and/or limestone powder-treatment of soil containing high sulfate level: ettringite minerals can hold a large amount of water within the material resulting in its expansion [14].

The untreated control soil sample exhibited the highest average volumetric expansion of 5.4% up to 10-day and then presented a decreasing trend of the expansion. This behavior may be attributed to the solubility of mineral compounds. At the beginning state of the expansion, the solubility of sulfate ions is high and dominates the volumetric expansion. But at a later age, the existence of sodium chloride reduces the solubility of sulfate ions and the crystallization of sodium sulfate is also reduced. Therefore, the expansion of the soil decreases eventually [15]. Moreover, the collapse of the soil layer structural system due to the ettringite formation under continuously moist conditions may also cause the reduction of the expansion after 10-day.

Generally, cement and limestone powder stabilization for sulfate-bearing soil is beneficial in terms of reducing the volumetric expansion of soil. Comparing treated samples, the soil samples stabilized with 2% and 4% limestone powder at fixed 4% cement experienced the relatively high expansion swelling, while the samples mixed with 6% cement and 6% with the addition of 2% limestone powder had the highest resistance against the volumetric increase.

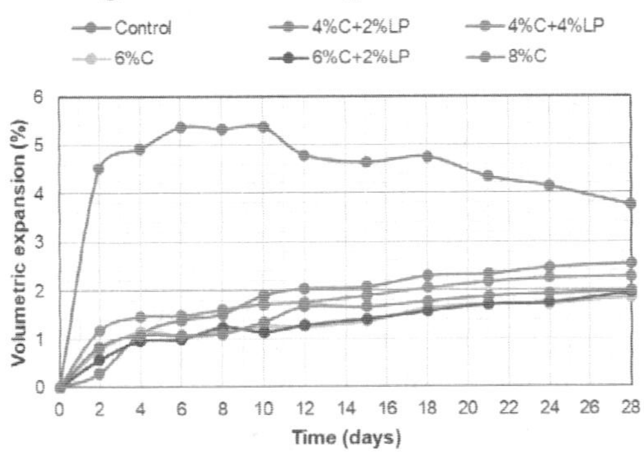

Figure 4. Three-dimensional swelling.

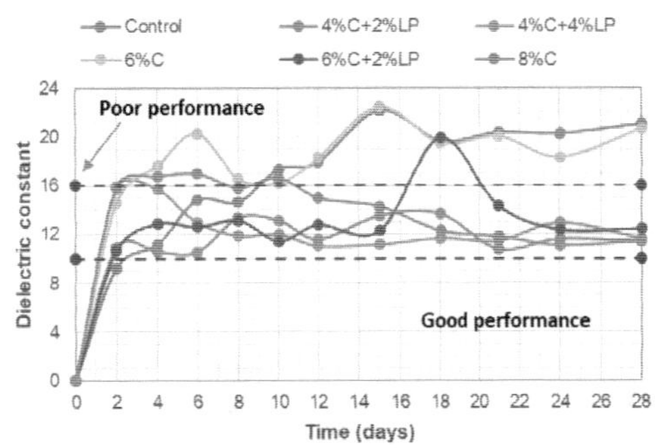

Figure 5. Dielectric constant.

3.5. Dielectric constant

Figure 5 shows dielectric constant (DC) values over time for the designed soil-cement-limestone powder mixtures. As stated earlier, the DC value represents the moisture susceptibility of soil material. The dielectric value is most sensitive and directly related to the amount of unbound water that exists within the soil mixture matrix [16]. The soil mixture with a final DC value less than 10 is considered as a good performance, while that with the DC value above 16 is expected to provide the poor performance in soil stabilization. The soil mixture having final DC values between 10 and 16 is expected to be marginally moisture-susceptible.

As with 3-D swelling results, there was a rapid increase in DC at an early age. Control soil mixture and the one treated with 6% cement exhibited higher DC values, 21.0 and 20.7 at 28-day, respectively, while the other mixtures stayed between 10 and 16 at the same age. The lowest DC value is observed in soil stabilized with a combination of 4% cement and 2% limestone powder. Interestingly, increasing limestone powder content leads to the increase of DC value, whereas the increase of cement content reduces the DC value. Therefore, it is important to select the optimum combination of cement and limestone powder contents to enhance the moisture susceptibility resistance of soil mixture.

4. Conclusion

In this paper, the combined effect of cement and limestone powder on geotechnical properties and durability of saline soil was investigated. In general, the stabilization of saline soil using the cement-limestone powder blend reduces sulfate-induced heaving and improve soil properties. The following conclusions can be drawn:

- The combination of cement and limestone powder (6% cement + 2 % limestone powder) is as effective in decreasing the Atterberg limits of soil as the cement used at higher content (8% cement).
- As stabilizer content increases, optimum moisture content increases and maximum dry density decreases.
- An increase in stabilizer content leads to strength improvement of sulfate-bearing saline soil, except 4% cement + 4% limestone powder blended mixture.
- The combined use of cement and limestone powder in the sulfate-bearing saline soil inhibited volumetric expansion due to ettringite minerals formed in the alumina-sulfate-calcium-water reaction.
- Cement and limestone powder stabilization increases the moisture susceptibility resistance of sulfate-bearing saline soil mixture.
- Based on the combined analyses of UCS, 3-D swelling, and DC results, the soil mixture stabilized with 6% cement and 2% limestone powder has the best performance for the mechanical and durability properties of sulfate-bearing saline soil.

Acknowledgments

This research was supported by the Nazarbayev University, School of Engineering and Digital Sciences. The authors are grateful for this support. Any opinions, findings, and conclusions or recommendations expressed in this material are those of the author(s) and do not necessarily reflect the views of the Nazarbayev University.

References

[1] Saylak, D., Mishra, S.K., Mejeoumov, G.G., Shon, C.-S 2008 Fly Ash-Calcium Chloride Stabilization in Road Construction *Transportation Research Record: Journal of the Transportation Research Board* **2053(1)** 23-29

[2] Cheng, Y., Wang, S., Li, J., Huang, X., Li, C., and Wu, J. 2018 Engineering and Mineralogical Properties of Stabilized Expansive Soil Compositing Lime and Natural Pozzolans *Construction and Building Materials* **187** 1031-1038

[3] Routhu, B. 2019 Stabilization of Soil by using Limestone Powder *International Research*

Journal of Engineering and Technology **6(3)** 6656-6661
[4] Al-Joulani 2012 Effect of Stone Powder and Lime on Strength, Compaction and CBR Properties of Fine Soils *Jordan Journal of Civil Engineering* **6(1)**
[5] Pastor, J., Tomás, R., Cano, M., Riquelme, A., & Gutiérrez, E. 2019 Evaluation of the Improvement Effect of Limestone Powder Waste in the Stabilization of Swelling Clayey Soil Sustainability **11(3)** 679 doi: 10.3390/su11030679
[6] AASHTO M 145-91 Standard Specification for Classification of Soils and Soil-Aggregate Mixtures for Highway Construction Purposes 2017 American Association of State Highway and Transportation Officials
[7] ASTM D4318-17 Standard Test Methods for Liquid Limit, Plastic Limit, and Plasticity Index of Soils 2018 Annual Book of ASTM Standards, ASTM International
[8] ASTM D698-12e2 Standard Test Methods for Laboratory Compaction Characteristics of Soil Using Standard Effort 2012 Standard Test Methods for Liquid Limit, Plastic Limit, and Plasticity Index of Soils. Annual Book of ASTM Standards, ASTM International
[9] Tex-121-E. Soil-Lime Testing 2002 Texas Department of Transportation
[10] Tex-120-E. Soil-Cement Testing 2013 Texas Department of Transportation
[11] ASTM D2166 Standard Test Method for Unconfined Compressive Strength of Cohesive Soil 2016 Annual Book of ASTM Standards, ASTM International
[12] Santos, F., Li, L., Li, Y., Amini, F 2011 Geotechnical Properties of Fly Ash and Soil Mixtures for Use in Highway Embankments. Proc., World of Coal Ash (WOCA) Conference, Denver 125-136
[13] Kumar, A., Walia, B.S., Bajaj, A. 2007 Influence of Fly Ash, Lime, and Polyester Fibers on Compaction and Strength Properties of Expansive Soil *Journal of Materials in Civil Engineering* **19-3** 242-248
[14] National Lime Association. 2000. Guidelines for stabilization of soils containing lime. Texas.
[15] Fang, F., Liu, J., Liu, C., Liu, Z., Ji., Y. 2018 The Crystallization and Salt expansion Characteristics of a Silty Clay *Cold Regions Science and Technology* **154** 63-73 https://doi.org/10.1016/j.coldregions.2018.06.009
[16] Shon, C.-S., Jung, Y.S., Saylak, D. 2013. Evaluation of Synthetic Aggregates Using Off-ASTM Specification Ashes as Road Base Course Materials *Construction and Building Materials* **38** 508-514 http://dx.doi.org/10.1016/j.conbuildmat.2012.08.014.

Surface electric field and adsorption properties of noble metal modified graphene in methane gas

Xinyu Huang, Weijian Li and Jijun Ding[1]

Shaanxi Key Laboratory of Measurement and Control Technology for Oil and Gas wells, College of Science, Xi'an Shiyou University, Xi'an, Shaanxi 710065, China

[1]E-mail: dingjj303@163.com

Abstract. Based on COMSOL Multiphysics software, the geometric model of graphene structure is set on the two-dimensional plane, and the thickness of graphene layer is set at 400 nm. The real and imaginary parts of the refractive index of graphene are defined. Then, the Ag and Au films are coated on both sides of graphene to form the Ag/G/Ag and Au/G/Au sandwich structures, respectively. In addition, the electric field modes between 300 nm and 600 nm are obtained by parameterized scanning calculation of the incident wavelength λ. At the same time, adsorption, scattering and extinction spectra are plotted. Results indicated that the maximum field strength of both the Ag/G/Ag and Au/G/Au sandwich structures are nearly doubled as compared to the intrinsic graphene. Finally, the adsorption, scattering and extinction properties are discussed. It is concluded that the resonance absorption peaks can be adjusted and improved by Ag/Au modification. The strongest plasma resonance mode shifts towards high energy. Moreover, Ag/G/Ag has stronger surface plasmon resonance effect than that of Au/G/Au structure.

1. Introduction

Graphene has been widely investigated due to its unique photoelectric properties. Such as, the structural, electrical evolution and optical properties of graphene induced by oxygen-related defects are investigated [1]. Lounis et al. [2] investigated the interaction between epitaxially grown graphene and SiC substrate using photoluminescence spectroscopy. Wu et al. [3] develop a high-efficiency technology to produce graphene paper with ultra-high thermal conductivity and large quantities. We also discussed the optical properties of graphene [4].

In addition, noble metal have strong local surface plasmon resonance (LSPR) characteristics related to the composition, morphology and size. Chen et al. [5] reported the LSPR of gold (Au) nanoparticles with different shapes and sizes, and investigated the response of their surface plasmon peaks to the refractive index of the surrounding medium. Ugwuoke et al. [6] presented a simplified model of plasmon hybridization and investigated the weak coupling regimes of solid and cavity plasmons in the core-shell nanostructures. Simone and Ruijter [7] demonstrated the surface plasmon resonance and surface-enhanced Raman scattering on a flat Au surface. In addition, plasmonic properties of Al-Ag alloys nanoparticles have been studied [8].

However, the application of graphene is limited due to intrinsic graphene is a gapless semiconductor. In this situation, previous literatures reported that noble metal modification can be used to improve the photoelectric properties of graphene. Lee et al. [9] assembled Au nanocrystals onto graphene and found that the surface enhanced Raman scattering depend on the morphologies of

Au nanocrystals. Zhao and Wu [10] concluded that noble metal (Rh, Pt, Pd) decorated graphene have higher adsorption ability, higher charge transfer and higher orbital hybridization than that of intrinsic graphene. Sharma and Rabinal [11] prepared graphene-silver nanoparticle composites for effective solar absorption by coupling surface plasmon and optical absorption of individual phases. Nair et al. [12] deposited Au on a monolayer graphene and proved that some of the gold are intercalated between the monolayer and the buffer layer.

Interestingly, noble metal modified graphene will produce coupling effect, which will improve the electric field distribution. This widens the resonant bandwidth of the semiconductor and expands the working range of noble metal LSPR. These provide the possibility to improve the detection sensitivity. Therefore, the noble metal modified graphene can be used as a candidate material for LSPR sensor with applications in environmental monitoring. For this reason, understanding of the plasma effect in noble metal modified graphene is of fundamental importance as well as of practical interest for application in sensor devices. Especially, silver is favored by researchers due to its small dielectric constant and suitable price. In this paper, a two-dimensional graphene plane model is established by COMSOL Multiphysics software, which replaces the thin layer structure of graphene. Then, the Ag and Au films are coated on both sides of graphene to form the Ag/G/Ag and Au/G/Au sandwich structures, respectively. The surface electric field of sandwich structures in methane gas are calculated, and their absorption, scattering and extinction spectra between 300 nm and 600 nm are plotted.

2. Model and method

Based on COMSOL Multiphysics software, the geometric model of graphene structure is set on the two dimensional plane, and the thickness of graphene layer is set at 400 nm. The refractive index of graphene is obtained from reference [13].

$$n_g = 3 + i\frac{C_1}{3}\lambda \tag{1}$$

Where, λ=400 nm is the wavelength of incident light, C_1=5.446 μm^{-1}, the real and imaginary parts of the refractive index of graphene are 3 and 0.72613, respectively. Then, the Ag and Au films are coated on both sides of graphene to form the Ag/G/Ag and Au/G/Au sandwich structures, respectively. Finally, the electric field modes between 300 nm and 600 nm are obtained by parameterized scanning calculation of the incident wavelength λ. At the same time, adsorption, scattering and extinction spectra are plotted.

3. Results and discussion

Figure 1 displays the schematic diagram of Ag/G/Ag and Au/G/Au sandwich structure. The thickness of graphene layer is set at 400 nm. The Ag and Au films are coated on both sides of graphene to form the Ag/G/Ag and Au/G/Au sandwich structures, respectively. The wavelength of incident light is λ=400 nm.

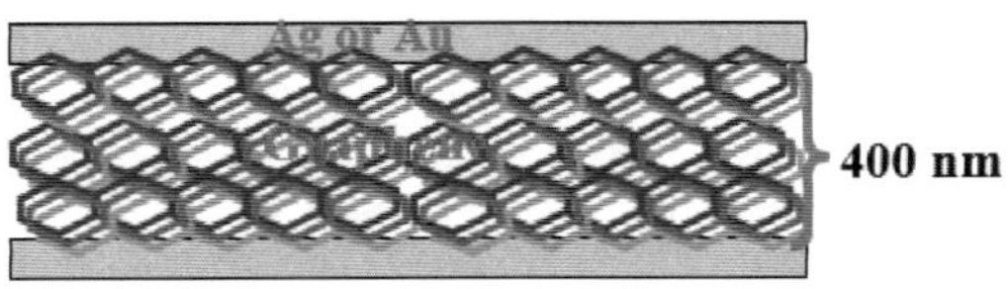

Figure 1. Schematic diagram of Ag/G/Ag and Au/G/Au sandwich structure.

Figure 2 shows surface electric field of noble metal modified graphene in methane gas: (a) intrinsic graphene, (b) Ag/G/Ag and (c) Au/G/Au. The Ag and Au films are coated on both sides of graphene to form the Ag/G/Ag and Au/G/Au sandwich structures, respectively. At λ=460 nm, the electric field enhancement can be observed on the right side of the intrinsic graphene. Compared with the intrinsic graphene, there is almost no electric field on both sides of graphene after modifying by Ag and Au films, while their maximum field strength is nearly doubled. This indicates that the sandwich structures have stronger absorption and scattering. In addition, the noble metal modification makes the surface and the nearby area of graphene have a strong near field enhancement. These provide the possibility to improve the detection sensitivity. Therefore, the noble metal modification graphene can be used as a candidate material for LSPR sensor with applications in environmental monitoring.

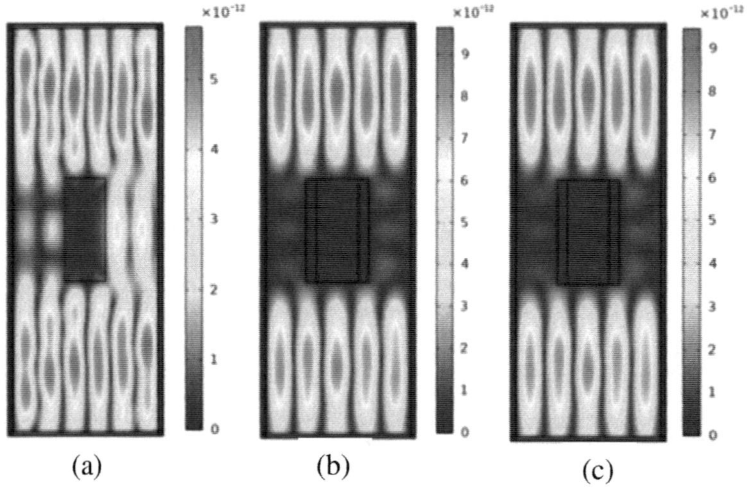

Figure 2. Surface electric field of noble metal modified graphene in methane gas: (a) intrinsic graphene, (b) Ag/G/Ag and (c) Au/G/Au.

Figure 3 presents adsorption spectra of noble metal modified graphene in methane gas: (a) intrinsic graphene, (b) Ag/G/Ag and (c) Au/G/Au. For intrinsic graphene, four adsorption peaks at 2.43 eV, 2.54 eV, 2.64 eV and 2.89 eV are observed. The intensity of the absorption peak attains the maximum at 2.64 eV. As Ag films are coated on both sides of graphene to form the Ag/G/Ag sandwich structures, all the adsorption peaks related to intrinsic graphene disappear, and the strongest absorption peak shifts towards high energy (3.11 eV). The results show that the position of resonance absorption peaks can be adjusted by Ag modification. However, there are four strong resonance absorption peaks in Au/G/Au structure, indicating that the surface plasmon resonance absorption is enhanced by Au modification.

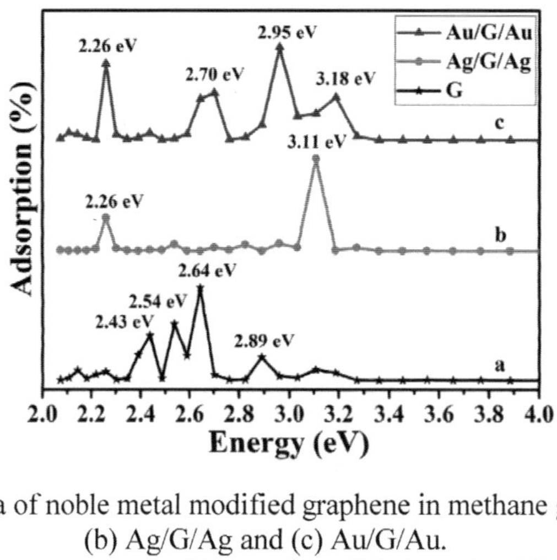

Figure 3. Adsorption spectra of noble metal modified graphene in methane gas: (a) intrinsic graphene, (b) Ag/G/Ag and (c) Au/G/Au.

Figure 4 displays extinction spectra of noble metal modified graphene in methane gas: (a) intrinsic graphene, (b) Ag/G/Ag and (c) Au/G/Au. The change of extinction peak is very sensitive to the modification of noble metal. The noble metals modification can obviously enhance the surface plasmon resonance of sandwich structures, which makes the extinction peak of Ag/G/Ag and Au/G/Au become stronger, and the strongest plasma resonance mode shifts towards high energy. Moreover, Ag/G/Ag has stronger surface plasmon resonance effect than that of Au/G/Au structure.

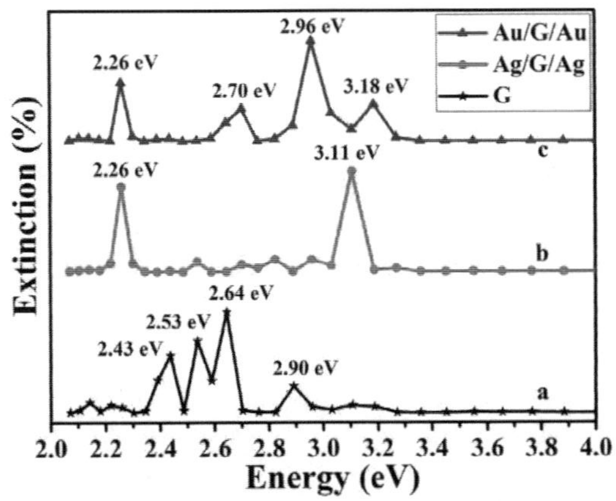

Figure 4. Extinction spectra of noble metal modified graphene in methane gas: (a) intrinsic graphene, (b) Ag/G/Ag and (c) Au/G/Au.

Finally, scattering spectra of noble metal modified graphene in methane gas are discussed, as shown in Figure 5. Compared with the absorption and extinction properties, the absorption of Au/G/Au is greater than that of scattering efficiency. On the contrary, Ag/G/Ag has higher scattering efficiency.

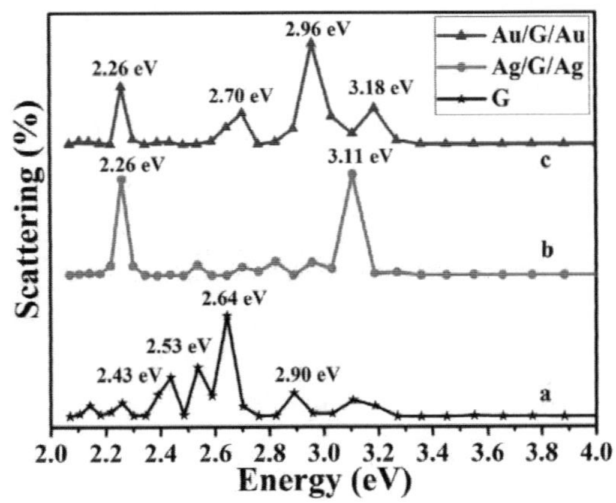

Figure 5. Scattering spectra of noble metal modified graphene in methane gas: (a) intrinsic graphene, (b) Ag/G/Ag and (c) Au/G/Au.

4. Conclusions

The surface electric field of intrinsic graphene, Ag/G/Ag and Au/G/Au sandwich structures are calculated by COMSOL Multiphysics software. Results indicated that the maximum field strength of both the Ag/G/Ag and Au/G/Au sandwich structures are nearly doubled as compared to the intrinsic graphene. In addition, adsorption, scattering and extinction spectra are discussed. It is concluded that the absorption peak intensity of intrinsic graphene attains the maximum at 2.64 eV. As Ag films are coated on both sides of graphene to form the Ag/G/Ag sandwich structures, all the adsorption peaks related to intrinsic graphene disappear, and the strongest absorption peak shifts towards high energy (3.11 eV). The results show that the position of resonance absorption peaks can be adjusted by Ag modification. However, there are four strong resonance absorption peaks in Au/G/Au structure, indicating that the surface plasmon resonance absorption is enhanced by Au modification. Moreover, Ag/G/Ag has stronger surface plasmon resonance effect than Au/G/Au structure. These results have theoretical and practical significance for the design of noble metal modified graphene optical sensors.

Acknowledgement
Authors thanks Professor Jie Wu of Xi'an Shiyou University for his help in software simulation calculation.

This work is supported by the Science and Technology Plan Program in Shaanxi Province of China (Grant No. 2019GY-170) and National Innovation Training Program for University Students in 2019 (Grant No. 201910705014).

References

[1] Wang S N, Wang R, Liu X F, Wang X W, Zhang D D, Guo Y J and Qiu X H 2012 Optical spectroscopy investigation of the structural and electrical evolution of controllably oxidized graphene by a solution method *J. Phys. Chem. C* **116** 10702-10707
[2] Lounis S D, Siegel D A, Broesler R, Hwang C G, Haller E E and Lanzara A 2010 Resonant photoluminescent charging of epitaxial graphene *Appl. Phys. Lett.* **96** 151913
[3] Wu T, Xu Y L, Wang H Y, Sun Z H, Zou L Y 2021 Efficient and inexpensive preparation of graphene laminated film with ultrahigh thermal conductivity *Carbon* **171** 639-645
[4] Ding J J, Chen H X, Feng D Q and Fu H W 2017 Investigation on photoluminescence emission of (reduced) graphene oxide paper *Mater. Sci. Eng.* **292** 012097
[5] Chen H J, Kou X S, Yang Z, Ni W H, Wang J F 2008 Shape and size-dependent refractive index sensitivity of gold nanoparticles *Langmuir* **24** 5233-5237
[6] Ugwuoke L C, Mančal T Krüger T P J 2020 Localized surface plasmon resonances of simple tunable plasmonic nanostructures *Plasmonics* **15** 189-200
[7] Simone G and Ruijter P 2020 Plasmon resonance excitation enhances Raman emission and amplifies the molecular vibration on Au (111) film *Appl. Surf. Sci.* **530** 147207
[8] Kuiri P K 2020 Tailoring localized surface plasmons in Ag-Al alloys' nanoparticles *J. Alloys Compd.* **826** 154250
[9] Lee Y H, Polavarapu L, Gao N, Yuan P and Xu Q H 2012 Enhanced optical properties of graphene oxide-Au nanocrystal composites *Langmuir* **28** 321-326
[10] Zhao C J and Wu H R 2018 A first-principles study on the interaction of biogas with noble metal (Rh, Pt, Pd) decorated nitrogen doped graphene as a gas sensor: A DFT study *Appl. Surf. Sci.* **435** 1199-1212
[11] Sharma B Rabinal M K 2017 Plasmon based metal-graphene nanocomposites for effective solar vaporization *J. Alloys Compd.* **690** 57-62
[12] Nair M N, Cranney M, Jiang T, Hajjar-Garreau S, Aubel D, Vonau F, Florentin A, Denys E, Bocquet M L and Simon L 2016 Noble-metal intercalation process leading to a protected adatom in a graphene hollow site *Phys. Rev. B* **94** 075427
[13] Bruna M, Borini S 2009 Optical constants of graphene layers in the visible range *Appl. Phys. Lett.* **94** 031901

Comparison of adsorbing NO_2 in boron and phosphorus doped graphene/ZnO heterojunction

Yanxin Jin, Jijun Ding[1] and Haixia Chen

College of Science, Xi'an Shiyou University, Xi'an, Shaanxi 710065, China

[1]Email: dingjj303@163.com

Abstract. As a new two-dimensional material, the adsorption properties of the graphene-like ZnO monolayer (ZnO-ML) have been investigated in our previous work. In this paper, the heterojunction are constructed using ZnO-ML and doped graphene. The electronic and adsorption properties of the boron doped graphene/ZnO (B-G/ZnO) and phosphorus doped graphene/ZnO (P-G/ZnO) heterojunction are investigated based on density functional theory (DFT). Firstly, the B-G/ZnO and P-G/ZnO heterojunction are constructed and their structures are optimized. Then, according to the comparison of adsorption energy and adsorption height, the most stable adsorption configurations are obtained. They are the T and T_O site corresponding to the B-G/ZnO and P-G/ZnO heterojunction, respectively. Finally, in order to further explore the adsorption mechanism between the heterojunction and NO_2 molecule, the band structure, density of states (DOS) and the partial density of states (PDOS) in B-G/ZnO and P-G/ZnO heterojunction are compared. These results can provide more theoretical support for the adsorption properties of heterojunction based on the ZnO-ML.

1. Introduction

With the development of the industry, more and more toxic and harmful gases are discharged into the atmosphere. Especially, the human health and environment are damaged by nitrogen oxide (NO_2) and carbon oxide (CO) [1]. Thus, it is necessary to investigate practical gas sensors to adsorb toxic gases. Graphene, due to its unique structure, can be used as a candidate material for gas sensors. It is successfully exfoliated from graphite in 2004 by Novoselov and Geim [2]. Graphene is a typical two-dimensional structure with the thickness of a carbon atom. In addition, it has good adsorption behaviors for gas molecule owing to a large specific surface area (2600 m^2/g) [3].

However, the pristine graphene has zero band gap. In order to overcome this problem, researchers devoted to open and control its band gap using heteroatom doping. The boron (B) and nitrogen (N) have similar radius as well as carbon atoms. Either B or N doping could change the band structure of graphene, which improves its electronic properties and adsorption ability for some gas molecules [4]. Thus, the B and N are usually selected as the doping atoms in graphene. Qin and Zhang [5] investigated that N doping graphene can enhance adsorption capacity for NO_2.

Interestingly, based on the density functional theory (DFT), monolayer zinc oxide (ZnO-ML) has more excellent adsorption properties than that of bulk ZnO. Besides, ZnO-ML has honeycomb structure like graphene. It is found that the ZnO-ML has excellent adsorption performance in our previous work [6]. Fu et al. [7] reported that graphene oxide and ZnO nanocomposites serving as a sensor exhibits a good repeatability and high sensitivity to ammonia. Chen et al. [8] found that NO_2 adsorbed on ZnO-ML with oxygen vacancy defects show strong adsorption energy.

Due to both graphene and ZnO-ML doped with heteroatoms having excellent adsorption characters, the heterojunction composed of them may have more interesting properties in adsorption. In this work, Firstly, the boron (B) and phosphorus (P) atoms are used to substitute carbon atoms in graphene, respectively. Then, the heterojunction are constructed using ZnO-ML and B/P doped graphene. The optimized structures of NO_2 adsorbed on the heterojunction are investigated by DFT, and the most stable geometry configurations are obtained. Finally, the adsorption energy, adsorption height, band structure, density of states (DOS) and partial density of states (PDOS) based on the most stable configuration are calculated. Results show that the NO_2 molecule adsorbed on both the G-terminal of B-G/ZnO heterojunction and ZnO-terminal of P-G/ZnO heterojunction have the largest adsorption energy than other cases.

2. Calculation method

All structural optimization calculations are carried out using DFT based on the Dmol^3 module [9], and the electron exchange-correlation is processed by the generalized gradient approximation (GGA) of Perdew-Burke Ernzerhof (PBE) [10]. It is supposed that the DFT semi-core pseudopotential (DSPP) and adopting double numerical plus polarization (DNP) basis set are propitious to study gas-molecule-adsorption systems. In order to correct van der Waals forces, the Grimme scheme is used [11]. And the vacuum layer is set as 20 Å to make the interaction between the periodic images reach the minimum. To ensure sufficient accuracy, the convergence tolerance of energy, max displacement and force are less than 1.0×10^{-5} Ha, 0.005 Å and 0.002 Ha/Å, respectively. For structure optimization, the global orbital cutoff is chosen as 4.4 Å in these computations, and the smearing energy is used to 0.005 Ha. The Brillouin zone is sampled with a 5×5×1 k-point grid with Gaussian smearing width of 0.2 eV. The initial model of graphene and ZnO-ML are derived from Materials Studio 8.0 package. The ZnO-ML is cleaved from (001) direction of bulk wurtzite. Then, the graphene layer (4×4×1 supercell) is introduced to match the 3×3×1 supercell of ZnO-ML. The lattice mismatch is 0.94%, which is similar with other literatures [12]. For heterojunction, the atoms are fully relaxed, the adsorption energy (E_{ad}) which is used to determine the stability of the adsorption model is calculated by using the following equation:

$$E_{ad} = E_t - \left(E_h + E_g\right) \quad (1)$$

Where, E_t is the total energy of adsorbing the gas molecules by the heterojunction, E_h and E_g are the total energies of the single heterojunction and the NO_2 gas molecule, respectively. In the heterojunction, the dependence of E_{ad} on charge state for each adsorption site is calculated to determine the most stable sites for adsorbing. The heterojunction have two adsorption terminals including graphene and ZnO. The graphene terminal have three typical adsorption sites, such as top (T), bridge (B) and hollow (H) sites, respectively. However, there are four adsorption sites in the ZnO terminal, and the T site is divided into the top of Zn (T_{Zn}) and O (T_O) atom, respectively.

3. Results and discussion

Different heterojunction models composed of ZnO-ML and doped graphene (B or P atoms) have been considered. The heterojunction are constructed after geometric optimization. Their formation energies are negative, which indicates that the models are stable and existing. In Figure 1 (a)-(d), there are two parts including upper and lower corresponding to the top and side views, respectively. Figure 1 (a)-(b) shows the optimized structures of B-G/ZnO and P-G/ZnO heterojunction. The initial distances between the layers are set as 3 Å in two kinds of heterojunction, after relaxation, they decrease to 2.86 Å and 2.06 Å respectively. It indicates that the P-G/ZnO has larger distortion than that of B-G/ZnO heterojunction, which is due to P having larger atom radius and stronger non-metallic properties than that of carbon atoms.

For each adsorbate, seven adsorption sites are considered, including three typical sites in the graphene terminal (T, B, and H) and four adsorption sites (T_{Zn}, T_O, B and H) in the ZnO terminal. In this work, NO_2 molecule is used as adsorption gas. According to equation (1), the adsorption energy

E_{ad} in seven adsorption sites are calculated, as shown in Table 1. All the adsorption energy is negative, indicating that it is an exothermic process and the reaction is spontaneous. For both the B-G/ZnO and P-G/ZnO heterojunction, as long as the terminal is the same, the adsorption energy is almost same for all the adsorption sites. The difference in adsorption energy between ZnO and graphene terminal is only about 0.2 eV for the B-G/ZnO heterojunction. Also, its adsorption energy is smaller than that of P-G/ZnO.

Figure 1 (c)-(d) shows the most stable adsorption configurations of NO_2 adsorbed on the B-G/ZnO and P-G/ZnO heterojunction after full relaxation, respectively. The initial distances between NO_2 and interface is also 3 Å. For B-G/ZnO, the T site of the graphene terminal has the largest adsorption energy (-0.797 eV) and shorter distance (d=1.76 Å) between the NO_2 and heterojunction. In addition, the NO_2 parallel to or perpendicular to the heterojunction is defined according to the angular relationship between the two nitrogen atom lines and the plane of the heterojunction. As shown in Figure 1 (c), NO_2 molecule is parallel to the heterojunction. No matter which the initial adsorption site is, T is the stable site. The introduction of NO_2 causes graphene distortion, and the B atom has a little angle in graphene. In the Figure 1 (d), the internal structure of the P-G/ZnO heterojunction have been squeezed by NO_2 molecules, and T_O is the most stable site with adsorption energy of -2.593 eV. The distance between the NO_2 and the ZnO interface is 1.72 Å, which is shorter than that of B-G/ZnO heterojunction due to P having stronger non-metallic properties than that of carbon atoms.

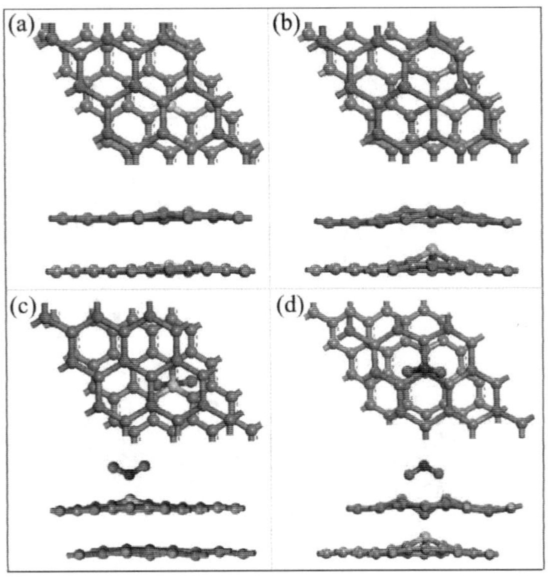

Figure 1. The most stable heterojunction configurations of (a) B-G/ZnO, (b) P-G/ZnO, (c) B-G/ZnO adsorbing NO_2 and (d) P-G/ZnO adsorbing NO_2.

Table 1. The adsorption energies of NO_2 adsorbed on different adsorption sites in heterojunction.

Adsorption Layer	B-G/ZnO		P-G/ZnO	
	Sites	E_{ad} (eV)	sites	E_{ad} (eV)
ZnO terminal	T_O/T_{Zn}	-0.577/-0.520	T_O/T_{Zn}	-2.593/-2.505
	B	-0.549	B	-2.590
	H	-0.554	H	-2.586
graphene terminal	T	-0.797	T	-0.587
	B	-0.792	B	-0.594
	H	-0.792	H	-0.587

For the P-G/ZnO heterojunction, the largest adsorption energies corresponding to T_O site of ZnO and B site of graphene terminal are -2.593 eV and-0.594 eV, respectively. Especially, the ZnO terminal adsorption energies of P-G/ZnO are larger, which is almost 5 times than that of B-G/ZnO. However, two kinds of heterojunction are belong to physical absorption, which indicates that there is no chemical bond between NO_2 molecule and heterojunction. For the B-G/ZnO, the most stable sites are T_O and H in the ZnO and graphene terminal, respectively. Also, the T_O and B are the most stable sites in the P-G/ZnO. A larger absolute value of adsorption energy correspond to a more stable adsorption configuration.

To understand the interaction between the doped graphene and ZnO-ML, the band structures of the heterojunctions are investigated, as can be seen in Figure 2 (a)-(b). The Fermi level lies on the zero energy and indicated by the black dashed line. The band gap could be modulated by doping and constructing heterojunction. As shown in Figure 2 (a), the band gap of B-G/ZnO heterojunction is 0.146 eV, and the band structure is only a superposition of B-G and ZnO-ML, it still retains the performance of doped graphene. The band structure of P-G/ZnO heterojunction is different from that of B-G/ZnO. In Figure 2 (b), there is an obvious flat band near the Fermi level, which indicates that the combination of the P-G and ZnO-ML introduces impurity states. Thus, the energy gap is increased to 0.57 eV. Also, the interaction between the NO_2 molecule and heterojunction are considered. In Figure 2 (c)-(d), the band structure of the NO_2 molecule adsorbed by B-G/ZnO and P-G/ZnO heterojunction are calculated, respectively. In Figure 2 (c), there is a flat band above the Fermi level, indicating the impurity is introduced due to the interaction between B-G/ZnO and NO_2. In Figure 2 (d), the flat band disappeared and the valence band is close to Fermi level. Meanwhile, the conduction band gradually approaches the Fermi level. The band gap of the NO_2 molecule adsorbed on B-G/ZnO and P-G/ZnO heterojunction are 0.556 eV and 0.331 eV, respectively. Compared with that before adsorption, the band gap of B-G/ZnO increases, but P-G/ZnO does the opposite. These results indicate that the heterojunction based on the ZnO-ML have great adsorption properties.

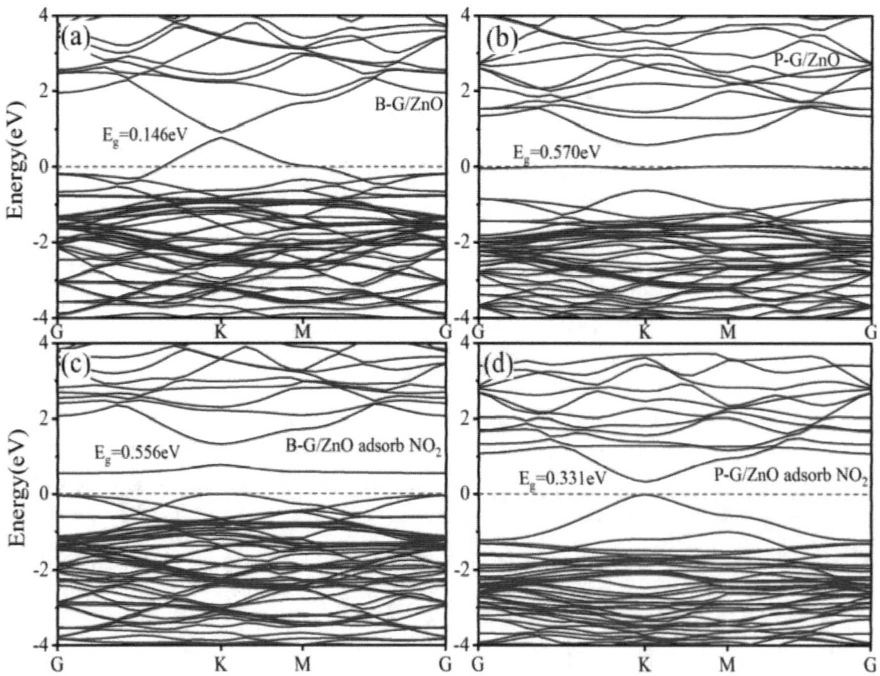

Figure 2. The band structures of (a) B-G/ZnO, (b) P-G/ZnO, (c) B-G/ZnO adsorbing NO_2 and (d) P-G/ZnO adsorbing NO_2, the Fermi level is set to zero energy and indicated by the red dashed line.

IOP Conf. Series: Materials Science and Engineering 1040 (2021) 012018 doi:10.1088/1757-899X/1040/1/012018

To further explore the adsorption mechanism, DOS and PDOS of B-G/ZnO and P-G/ZnO heterojunction before and after adsorbing NO_2 molecule are calculated, as displayed in Figure 3. The PDOS of the B, C, N, O and P atoms' electron orbitals are analyzed. For B-G/ZnO heterojunction, the number and the maximum of peaks are almost unchanged (Figure 3a and 3c). After adsorbing NO_2, there are two weak peak with high energy state near the Fermi level, which is consistent with its band structure. It is explained that the impurity is introduced. The B-G/ZnO heterojunction has a maximum peak at -5.1 eV, which is mainly contributed by *d* orbital of Zn atom. The peak of the high energy state is contributed by the *p* orbital of B atom. After adsorbing NO_2, it is observed that the *s* and *p* orbitals of O atom affect the small peak near the Fermi level and low energy state in total DOS. Besides, the other orbitals are similar with that before adsorption. For the P-G/ZnO heterojunction, the small peak disappear near the Fermi level, at the same time, there is a new weak peak in -21.4 eV after adsorption. It is mainly contributed by the *s* orbital of O atom, *s*, *p* and *d* orbitals of P atom, respectively. It is proved that graphene is effectively doped. Besides, there appear some new peaks, including the peak near -12.1 eV attributed to *s* orbitals of O and P atoms, and the peak near -9.3 eV belonged to *p* orbital of O atom. The maximum peak in B-G/ZnO and P-G/ZnO heterojunction are different, one is in the right of -5 eV, and the other is in the left of -5 eV. However, the maximum peak are almost same. In summary, due to a number of great electrical properties, B-G/ZnO and P-G/ZnO heterojunction can be considered as excellent gas sensitive materials.

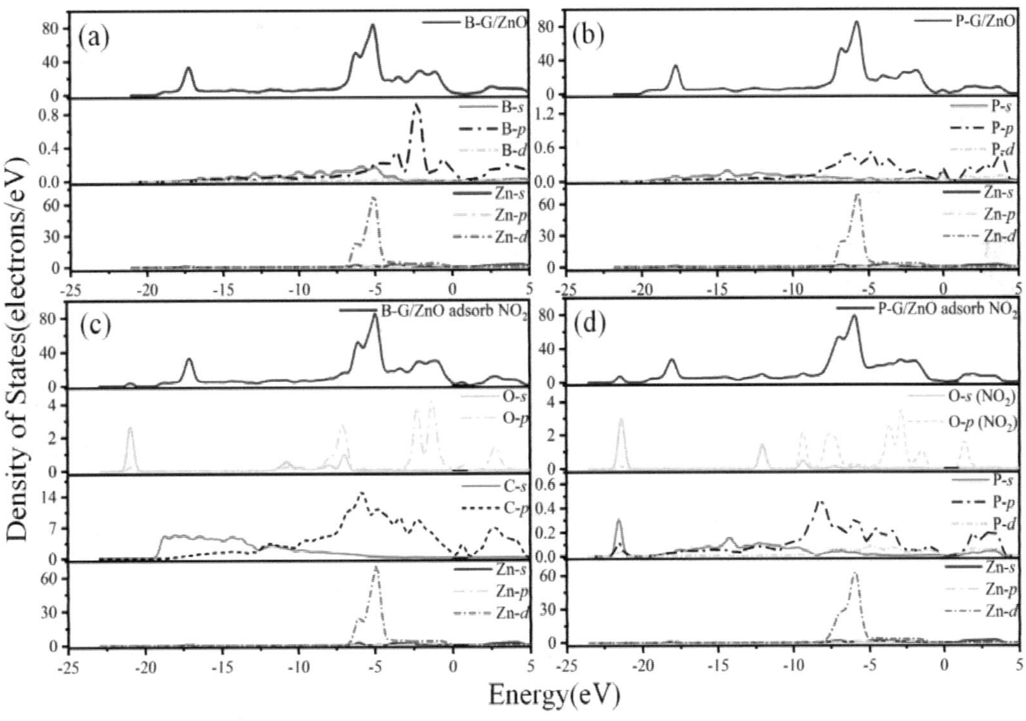

Figure 3. The DOS and PDOS of (a) B-G/ZnO, (b) P-G/ZnO, (c) B-G/ZnO adsorbing NO_2 and (d) P-G/ZnO adsorbing NO_2 based on the most stable configurations.

4. Conclusion

The adsorption and electronic properties of B-G/ZnO and P-G/ZnO heterojunction with the most stable configurations are investigated using the DFT. The most stable sites that NO_2 molecules can be effectively absorbed are the T site of B-G/ZnO heterojunction and the T_O site of P-G/ZnO heterojunction, respectively. They have the largest adsorption energies and the shortest adsorption heights than other cases. The adsorption capacity of pristine graphene can be greatly enhanced by doping and constructing heterojunction. It implies that B/P atom doping can modulate the band gap and the electronic structure more flexibly. Finally, these results can provide more theoretical support for the adsorption properties of heterojunction based on the ZnO-ML, which could become a promising gas sensitive materials.

Acknowledgement
Authors thanks Professor Jianhong Peng of Qinghai Nationalities University for his help in software simulation calculation.

This work is supported by the National Natural Science Foundations of China (Grant No. 11804273; 11447116), Science and Technology Plan Program in Shaanxi Province of China (Grant No. 2019GY-170; 2016JQ5037), Special Program for Scientific Research of Shaanxi Educational Committee (Grant No. 16JK1601), Graduate Student Innovative and Practical Ability Training Program of Xi'an Shiyou University (Grant No. YCS19211036).

References

[1] Esrafili M D, 2019 Boron and nitrogen co-doped graphene nanosheets for NO and NO_2 gas sensing *Phys. Lett. A* **383** 1607-1614
[2] Novoselov K S, Geim A K, Morozov S V, Jiang D, Zhang Y, Dubonos S V, Grigorieva I V and Firsov A A, 2004 Electric field effect in atomically thin carbon films *Science* **306** 666-669
[3] Dsouza R, Vats T, Chattree A and Siril P F, 2018 Graphene supported magnetically separable solid acid catalyst for the single step conversion of waste cooking oil to biodiesel *Renew. Energ.* **126** 1064-1073
[4] Tian W C, Liu X H and Yu W B, 2018 Research progress of gas sensor based on graphene and its derivatives: A Review *Appl. Sci.* **8** 1118
[5] Qin Y X and Zhang Z, 2020 Co-modulation of graphene by N hetero-doping &vacancy defects and the effect on NO_2 adsorption and sensing: First-principles study *Physica E* **116** 113737
[6] Qu Y F, Ding J J, Fu H W, Peng J H and Chen H X, 2020 Adsorption of CO, NO, and NH_3 on ZnO monolayer decorated with noble metal (Ag, Au) *Appl. Surf. Sci.* **508** 145202
[7] Fu H W, Jiang Y H, Ding J J, Zhang J L, Zhang M, Zhu Y and Li H D, 2018 Zinc oxide nanoparticle incorporated graphene oxide as sensing coating for interferometric optical microfiber for ammonia gas detection *Sens. Actuators, B* **254** 239-247
[8] Chen H X, Qu Y F, Ding J J and Fu H W, 2019 Adsorption behavior of graphene-like ZnO monolayer with oxygen vacancy defects for NO_2: a DFT study *Superlattices Microstruct.* **134** 106223
[9] Delley B, 1995 DMol, a standard tool for density functional calculations: review and advances *Theor. Comput. Chem* **2** 221-254
[10] Perdew J P, Burke K and Ernzerhof M, 1996 Generalized gradient approximation made simple *Phys. Rev. Lett.* **77** 3865-3868
[11] Grimme S, 2006 Semi empirical GGA-Type Density Functional constructed with a long-range dispersion correction *J. Comput. Chem.* **27** 1787-1799
[12] Gholizadeh R, Yu Y X, Wang Y J, 2017 N_2O adsorption and decomposition over ZnO (0001) doped graphene: Density Functional Theory calculations *Appl. Surf. Sci.* **420** 944-953

AUTHOR INDEX